THE NATURAL WORLD

TimeFrame

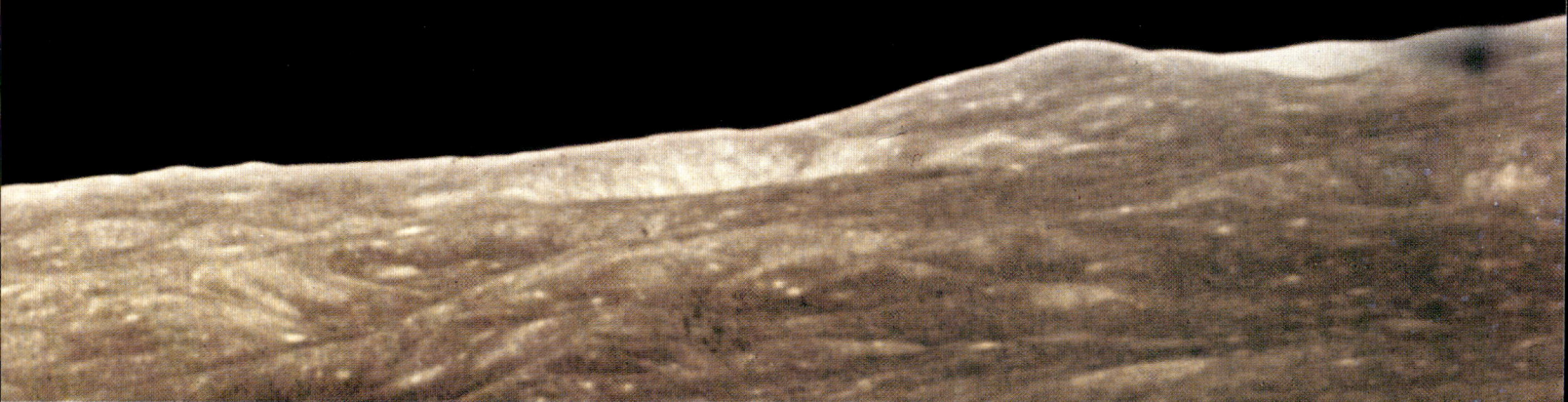

TIME® LIFE BOOKS

Other Publications:
TIME-LIFE LIBRARY OF CURIOUS AND UNUSUAL FACTS
AMERICAN COUNTRY
VOYAGE THROUGH THE UNIVERSE
THE THIRD REICH
THE TIME-LIFE GARDENER'S GUIDE
MYSTERIES OF THE UNKNOWN
FIX IT YOURSELF
FITNESS, HEALTH & NUTRITION
SUCCESSFUL PARENTING
HEALTHY HOME COOKING
UNDERSTANDING COMPUTERS
LIBRARY OF NATIONS
THE ENCHANTED WORLD
THE KODAK LIBRARY OF CREATIVE PHOTOGRAPHY
GREAT MEALS IN MINUTES
THE CIVIL WAR
PLANET EARTH
COLLECTOR'S LIBRARY OF THE CIVIL WAR
THE EPIC OF FLIGHT
THE GOOD COOK
WORLD WAR II
HOME REPAIR AND IMPROVEMENT
THE OLD WEST

For information on and a full description of
any of the Time-Life Books series listed above,
please call 1-800-621-7026 or write:
Reader Information
Time-Life Customer Service
P.O. Box C-32068
Richmond, Virginia 23261-2068

This volume is one in a series that tells the story
of humankind. Other books in the series include:
The Human Dawn
The Age of God-Kings
Barbarian Tides
A Soaring Spirit
Empires Ascendant
Empires Besieged
The March of Islam
Fury of the Northmen
Light in the East
The Divine Campaigns
The Mongol Conquests
The Age of Calamity
Voyages of Discovery
The European Emergence
Powers of the Crown
Winds of Revolution
The Pulse of Enterprise
The Colonial Overlords
The World in Arms
Shadow of the Dictators
The Nuclear Age
The Rise of Cities

THE NATURAL WORLD
TimeFrame

BY THE EDITORS OF TIME-LIFE BOOKS
TIME-LIFE BOOKS, ALEXANDRIA, VIRGINIA

Time-Life Books is a division of Time Life Inc., a wholly owned subsidiary of
THE TIME INC. BOOK COMPANY

TIME-LIFE BOOKS

Managing Editor: Thomas H. Flaherty
Director of Editorial Resources:
Elise D. Ritter-Clough
Director of Photography and Research:
John Conrad Weiser
Editorial Board: Dale M. Brown, Roberta Conlan, Laura Foreman, Lee Hassig, Jim Hicks, Blaine Marshall, Rita Thievon Mullin, Henry Woodhead

PUBLISHER: Joseph J. Ward

Associate Publisher: Ann Mirabito
Editorial Director: Russell B. Adams
Marketing Director: Anne Everhart
Director of Design: Louis Klein
Production Manager: Prudence G. Harris
Supervisor of Quality Control:
James King

EUROPEAN EDITOR: Ellen Phillips
Design Director: Ed Skyner
Director of Editorial Resources:
Samantha Hill
Chief Sub-Editor: Ilse Gray
Assistant Design Director: Mary Staples

Correspondents: Elisabeth Kraemer-Singh (Bonn); Maria Vincenza Aloisi (Paris); Ann Natanson (Rome). Valuable assistance was also provided by: Louise D. Forstall (Alexandria, Virginia); Yang Jinquan (Beijing); Elizabeth Brown (New York); Josephine du Brusle (Paris); Traudl Lessing (Vienna).

TIME FRAME
(published in Britain as
TIME-LIFE HISTORY OF THE WORLD)
SERIES EDITOR: Charles Boyle

Editorial Staff for *The Natural World*
Editor: Charles Boyle
Designer: Lynne Brown
Researchers: Lesley Coleman, Celia Dearing, Deborah Pownall
Sub-Editors: Frances Willard,
Luci Collings, Tim Cooke
Design Assistant: Sandra Archer
Editorial Assistant: Molly Sutherland

Picture Department
Picture Administrator: Amanda Hindley
Picture Coordinator: Elizabeth Turner

Editorial Production
Production Assistant: Emma Veys
Editorial Department: Theresa John,
Debra Lelliott, Juliet Lloyd-Price

U.S. EDITION

Assistant Editor: Barbara Fairchild Quarmby
Copy Coordinator: Ann Lee Bruen
Picture Coordinator: David Beard

Editorial Operations
Production: Celia Beattie
Library: Louise D. Forstall
Computer Composition: Deborah G. Tait (Manager), Monika D. Thayer, Janet Barnes Syring, Lillian Daniels

Special Contributors: Neil Fairbairn, Ellen Galford, Michael Kerrigan, Alan Lothian, Clio Whittaker (text); David E. Manley (index).

CONSULTANTS

General:
NEIL C. ROBERTS, Lecturer in Physical Geography, Loughborough University of Technology, England

Classical:
GRAEME BARKER, Professor of Archaeology, University of Leicester, England

China:
CHRISTOPHER CULLEN, Lecturer in the History of Asian Technology, School of Oriental and African Studies, University of London

Medieval and Early Modern:
DENIS COSGROVE, Reader in Cultural Geography, Loughborough University of Technology, England

Library of Congress Cataloging in Publication Data

The Natural world / by the editors of Time-Life Books.
      p.   cm. — (Time frame)
   Includes bibliographical references and index.
   ISBN 0-8094-6491-8
   ISBN 0-8094-6492-6 (lib. bdg.)
   1. World history. 2. Man—Influence on nature. 3. Man—Influence of environment.
   I. Time-Life Books. II. Series: Time frame.
D21.3.N38 1991
909—dc20   91-241
                                                CIP

© 1991 Time-Life Books. All rights reserved. No part of this book may be reproduced in any form or by any electronic or mechanical means, including information storage and retrieval devices or systems, without prior written permission from the publisher, except that brief passages may be quoted for reviews.
First printing. Printed in U.S.A.
Published simultaneously in Canada.
School and library distribution by Silver Burdett Company, Morristown, New Jersey 07960.

TIME-LIFE is a trademark of Time Warner Inc. U.S.A.

Time Life Inc. offers a wide range of fine recordings, including a *Rock 'n' Roll Era* series. For subscription information, call 1-800-621-7026 or write Time-Life Music, P.O. Box C-32068, Richmond, Virginia 23261-2068.

CONTENTS

1 **Out of the Wilderness** 8
Essay: Sanctuaries of the Spirit   31

2 **The Mediterranean World** 38

3 **The Chinese Way** 58
Essay: Mining the Earth's Riches   79

4 **The Plow and the Cross** 86

5 **Expansion and Exchange** 110
Essay: The Odysseys of Plants   133

6 **The Global Challenge** 140

Chronology   166
Picture Credits   168
Bibliography   169
Acknowledgments   171
Index   171

OUT OF THE WILDERNESS

Trekking across the desert wastes of the Rub al-Khali of Saudi Arabia, a party of Bedouin nomads accompanied by an English explorer advanced in silence under the sun's relentless glare. They progressed at a steady walking pace in order to conserve the energy of their camels: The only food available for these beasts consisted of the dry leaves and thorns of occasional bushes, and it might be several days before the party reached the next water hole. On the flat gravel plains, they averaged around three miles an hour; where the dunes were steep and the sands yielded at every footstep, they managed no more than half this rate. They had seen no other human beings for seventeen days.

Abruptly, the party came to a halt on hard ground. A gray-bearded Bedouin dismounted and examined some camel tracks that were partly blurred by the wind. "Who were they?" demanded the group's leader. The bearded man broke some camel droppings between his fingers and reported that the tracks had been made ten days earlier by six members of a particular tribe; the riders had raided another tribe on the southern coast and had stolen three of their camels; and they had last taken water at a well named Mughshin.

The Englishman Wilfred Thesiger had learned not to be surprised by his companions. "Here every man knew the individual tracks of his own camels," he reported, "and some of them could remember the tracks of nearly every camel they had seen. They could tell at a glance from the depth of the footprints whether a camel was ridden or free, and whether it was in calf. By studying strange tracks they could tell the area from which a camel came."

For the Bedouin, reading imprints in the sand with such accuracy was as commonplace a skill as is, say, a contemporary city dweller's ability to drive a car or purchase a theater ticket by telephone. This skill was also crucial to their survival: It could tell them the whereabouts of their enemies, a precious water hole, or an animal for the cooking pot. And this was just one example of their intimate cognizance of the harsh environment in which they lived. Their hereditary knowledge had enabled them to feed and clothe themselves and cure their ailments by methods that had hardly varied for thousands of years.

Wilfred Thesiger traveled with the Bedouin for many months in the late 1940s, and the qualities he came to admire most in his companions were their dignity, courage, endurance, and generosity. But when in the 1970s the oil wealth of Saudi Arabia transformed forever the lives of the Bedouin, when trucks replaced camels, and television antennas sprouted from black felt tents, what the rest of the world had most to mourn was the loss of an entire memory bank of knowledge of the natural world. When skills are no longer needed, they quickly die out.

Three thousand years ago, the knowledge possessed by the Bedouin was akin to

Painted some 4,000 years ago on a rockface in the region of Tassili N' Ajjer—or "Plateau of the Rivers"—in southern Algeria, a hunter launches his spear toward a horned wild sheep. Snapping at the heels of the prey is a dog, probably the first animal species to be domesticated by humans. Between 7000 and 3000 BC, much of the now-arid Sahara was well-watered grassland inhabited by humans who gathered edible plants and hunted herds of wild game, including antelope, giraffes, ostriches, and elephants. Some regions also yielded freshwater turtles, mollusks, and fish.

Amid the trees of the Kalahari Desert in southern Africa, crouching hunters of the !Kung people keep downwind of their prey. Their bows and arrows hardly differ from those used by their ancestors around 10,000 BC. The spears they carry to finish off a wounded animal have an even longer pedigree: Flaked points such as the 40,000-year-old example on the right—which is shaped to slot into a shaft of wood or bone—were fashioned beginning about 125,000 BC, enabling hunters to kill with new efficiency.

that of all other human communities inhabiting the earth. Most of these people lived by one of three overall strategies: They survived by hunting animals and gathering plants, by simple methods of agriculture, or—like the Bedouin—off herds of domesticated animals migrating across huge tracts of land in search of fresh pastures. Each of these lifestyles originally evolved in response to particular environmental conditions, and for countless generations, they proved capable of satisfying all basic human needs for food and shelter.

Later and more populous societies were to become more exacting in their needs, more specialized in their use of natural resources, and more vulnerable to crises. Successive civilizations inhabiting the Mediterranean region, culminating in the awesome grandeur and power of the Roman Empire, discovered that the progressively more efficient exploitation of nature can reap diminishing returns. In China and other parts of Asia, the needs of fast-growing human populations tested the limits of human skill and ingenuity. The Western world, moving by degrees from levels of bare subsistence to unparalleled affluence, solved its own problems of food supply by the application of advanced scientific and technological processes—and came late to a realization that short-term gains can be offset by long-term losses on an immeasurably greater scale. But the experience of humans who lived before cities were invented has never lost its relevance, for they too were not immune to ecological crises, and the pattern of their development followed a now-familiar course.

Even as they roamed vast areas taking what nature offered—hunting mammals and birds, catching fish, and gathering nuts, berries, and other plant foods—the earliest humans were beginning to shape the development of the world they lived in. Having come to know the ways of this world so intimately that they could anticipate where and when a particular plant could be expected to yield its richest harvest, and where migrating herds might most easily be ambushed, they proceeded to intervene directly in the world's workings—by burning wooded land, for instance, to encourage the growth of favored food plants. It was only a matter of time before humans began to cultivate plants as the foundation of their economic existence. Settled in one place and growing their own food, they were able to accumulate a surplus with which they could attract and tame wild grazing animals—sources of meat, wool, hide, and haulage. In some drier regions, communities that proved successful in stock raising eventually took to the nomadic life once more—this time as pastoralists, driving their herds from one area to another, often over great distances, to take full advantage of the seasonal growth of pastures.

Such specialized lifestyles provided sustenance, and at times even affluence. Yet these achievements were bought at a high price: The more control humans gained over their environment, the more serious and lasting was the impact they had on it; at the same time, the more they came to depend on single sets of resources, abandoning the opportunistic flexibility of the earliest hunter-gatherers, the more immediate and severe was the impact of any environmental change upon them.

Fortunately, because potentially of great benefit to the human understanding of the natural world, there still exist in remote and inhospitable parts of the globe peoples who maintain a traditional relationship with their habitat that is relatively undisturbed by the political and economic imperatives of modern civilization. These people represent a minute proportion of the global population; but their lifestyles are palpable evidence of how all humanity once lived, and of a closeness to nature that most of humankind has lost.

The world into which humans first emerged was itself in the grip of profound and continuous change. The shape of the land was being molded by the effects of a succession of great freezes, each lasting for about 100,000 years, which were interrupted by shorter, milder periods. Massive ice sheets crept southward century by century, covering what is now Canada and Scandinavia; at times as much as three miles thick, they carved out new landscapes as they scraped and gouged their way across the northern continents. Both northern and southern ice sheets drew moisture from the earth's atmosphere; as a result, sea levels fluctuated sharply, dropping by more than 300 feet during the most frigid periods and redrawing the map of the world. Asia and North America, separated in modern times by the Bering Strait, were joined by a wide, low-lying land bridge across which both animals and humans were able to make their way. Far to the west, meanwhile, the British Isles, along with a large area of the North Sea, formed a continuous promontory of continental Europe.

While the homelands of the first human beings in southern Africa remained subtropical through even the most frigid periods, the Eurasian areas that modern humans began to settle some 40,000 years ago were bearing the brunt of the great glaciations. There, where the sterile expanse of the ice sheet ended, the tundra began—a vast, monotonous terrain that no summer thawed beyond the top few inches. Farther south, a broad belt of grassland or steppe stretched across the breadth of Eurasia. Burgeoning in a blaze of color through the few short weeks of summer each year, its millions of square miles of grass, sparse and tussocky as it was, supported large herds of grazing mammals—reindeer, horses, mammoths, woolly rhinoceroses, and muskoxen. Beyond the steppe lay forest refuges; and from here, as the ice sheets receded during milder periods, leafy deciduous woodland pushed northward.

In western Europe, warmed by the deep waters of the Atlantic Ocean, the winters were milder and the summers cooler than those at the heart of the Eurasian landmass. Between 30,000 and 12,000 years ago, this area was home to large numbers of humans who took meat from the rich game reserves the continent offered and gathered vegetable food from a wide range of plants. Living mostly in small bands of around twenty-five people, they would regularly come together into larger tribal groups to mount communal hunts or to sit out the dark months of winter in caves. During the spring and summer months, on the other hand, when the blooming steppe provided an abundance of easily gathered plant foods, these bands might fragment into smaller family units comprising simply a man, a woman, and their children.

The size of the social group was very closely determined by the available food supply. Its members cooperated closely, sharing both food and leadership, though by and large operating a very clear sexual division of labor. Men hunted animal foods; women gathered vegetable products—nuts, berries, edible leaves, and grasses—as well as birds' eggs, small mammals, and by rivers and coasts, shellfish.

Roaming at any given time in a relatively confined tract of land of a few hundred square miles, within a much larger territory of up to 1,600 square miles, a hunter-gatherer band would periodically venture farther afield to occupy a new tract. Each area of land was thus given a chance to recover after a few years' exploitation, while the group remained fa-

A rock painting in Spain shows angry bees swarming around a precariously balanced woman as she reaches into their nest to extract honey. Among the foodstuffs gathered by women—including plants, roots, berries, insects, and birds' eggs—honey was a prized delicacy and was eaten together with the larvae and the comb. Some contemporary hunter-gatherers make a mildly alcoholic wine by squeezing the honey from the comb, dissolving it in warm water and adding yeast, then leaving it to ferment for a day and a night.

miliar with the full range of resources offered by its territory as a whole without becoming overdependent on the products of one particular zone. Dispersed in this way over hundreds of square miles, the hunter-gatherer band kept its impact on its environment to a minimum. The impact was, for all that, by no means negligible.

During colder periods, the dietary mainstay of the early European hunter-gatherers was provided by the herds of large mammals such as bison, red deer, horses, reindeer, and mammoths that grazed the open steppe. Milder periods saw an increase in the importance of smaller, woodland species such as roe deer and wild pigs. Hunter-gatherers living in warmer climates tended to rely more heavily on vegetable foods, which in the tropics were abundant throughout the year and easy to find, if laborious to gather. But the growing season in western Europe at cold times was very brief, and a group's women had to be able to respond extremely quickly and with careful planning if they were to exploit its harvest to the full. The important plant varieties would be scattered in different habitats across a wide area, and for each to be caught at its moment of ripening required careful organization.

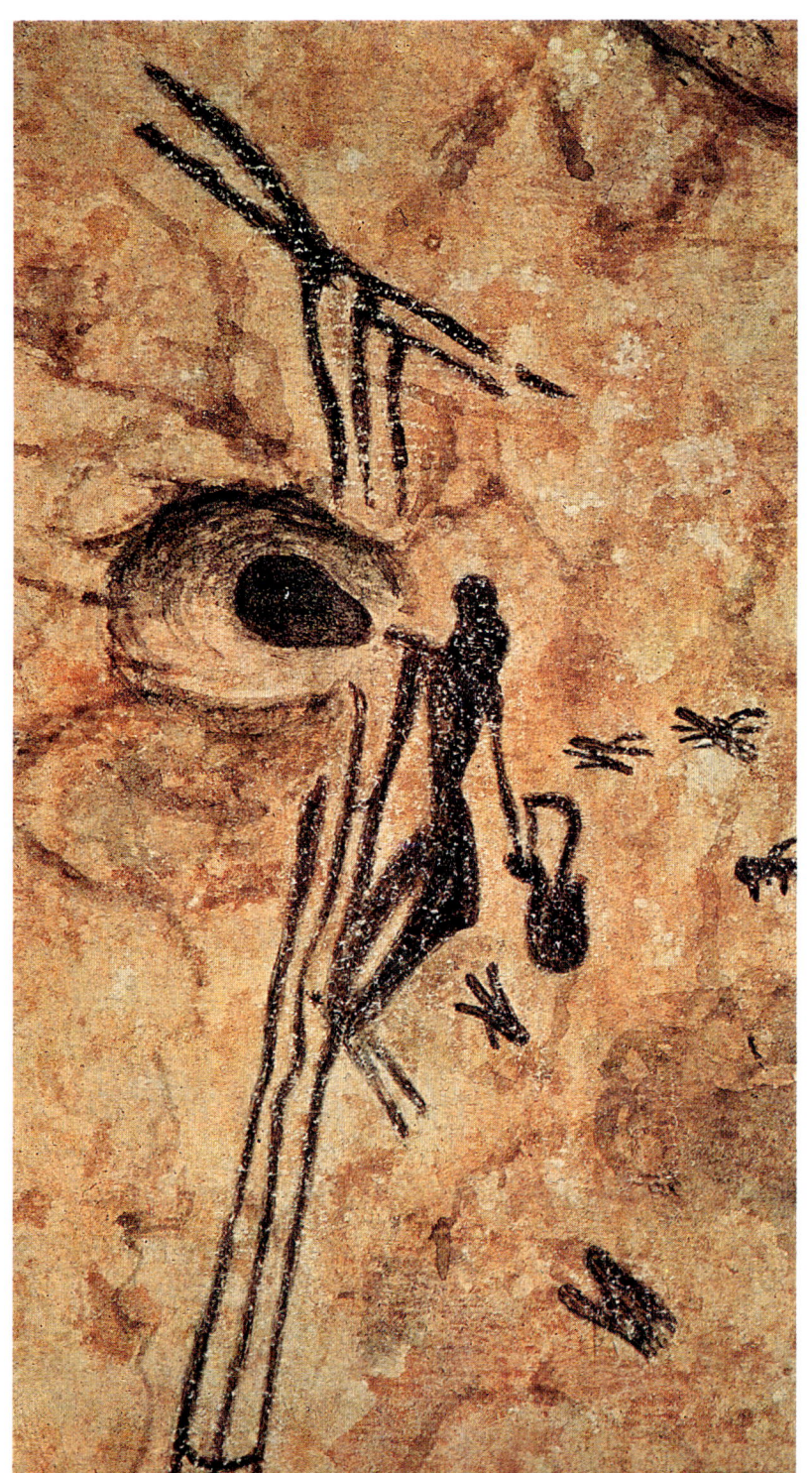

Moreover, the considerable nutritional value offered by plant foods and such animal foods as birds' eggs and shellfish was to some degree offset by the labor required in their gathering and preparation. While it might take a group of hunters a good deal of time, ingenuity, and luck to hunt down a red deer, for example, that animal once killed would provide the caloric equivalent of more than 52,000 oysters, each of which would need laborious picking and shelling before its nutritional possibilities could be realized. Given the abundance of grazing animals the hunter-gatherers had access to, it was inevitable that they would tend to rely greatly on meat in their diet. Animals provided a range of other products, too: from hide, fur, and sinew for making blankets and sewn clothing, to bone and ivory for tools, weapons, and—where wood was scarce—for use in bulk as a building material and fuel.

Around 20,000 years ago, humans developed technologies that made hunting significantly easier and less dangerous than it had been. For earlier hunters, armed only with cumbersome stone clubs and spears and crude axes, the hunt had been a rough and grueling battle between man and beast, fought at close quarters and at considerable risk to the man. Now, however, the bow and arrow were introduced, and the spear-thrower was added to the hunter's arsenal. A length of wood or bone about one and one-half feet long, with a hook at one end that clipped over the butt end of a spear, the spear-thrower acted as an extension of the human arm:

Held back over the shoulder, then snapped sharply forward, it sent the spear flying swiftly through the air toward its target, straight and true. No longer a crude stabbing weapon, the spear became a deadly missile, accurate up to about 100 feet. Fine stone blades, mass-produced by expert craftsmen from prepared flint cores, and delicate slivers of bone cunningly fashioned into lethal points added to the impact of both arrows and spears. Hunting was becoming less a dangerous, physical brawl between hunters and hunted and more an art that depended on skill and finesse: a battle of wits in which man's advantages were clear.

Along with the new weapons came changes in tactics. Cleverly planned ambushes and game drives now allowed hunters to kill in huge numbers, making them far more dangerous than any animal predator had ever been. Siting their encampments near river fords, the bands of hunters positioned themselves to attack herds of migratory mammals at a vulnerable moment during their annual journey. Mammoths might be ambushed where a gently sloping hillside offered a convenient butchering floor on which the carcasses—far too heavy to be dragged about—could be carved up where they fell, and on which the meat obtained could then be dried in the sun.

Fire, which was first used by humans to scare off predatory animals and to cook food, was also employed in hunting. Not all fires were started by humans; many of them must have occurred naturally, caused by bolts of lightning. But the deliberate use of fire in hunting was sufficiently widespread to have had a major impact on the environment. In some regions, when a patch of woodland or scrub was burned, the extra sunlight and the nutrient-rich layer of ash that resulted at ground level encouraged the growth of food plants—whether for humans to eat themselves or as a means of attracting game animals. On the other hand, the burning of trees and undergrowth left topsoil unprotected and liable to be washed away by subsequent rain showers, especially on sloping ground.

THE DIVINE RADIANCE

Until the world's resources were manipulated to yield other sources of energy, all human beings were utterly slaves to the sun, the foundation of every form of life that they encountered. The growth, survival, and reproduction of plants are dependent on the sun's radiant energy, which is trapped by their green leaves; when plants are eaten by other creatures, the same energy powers movement and the growth of body tissues. The images on these pages and overleaf illustrate the high position accorded to the sun in the pantheons of early societies—an intuitive acknowledgment of the debt that their members owed to the golden disk that swung across their skies, and whose nightly vanishing intimated their mortality.

On a stele of the ninth century BC, the Mesopotamian sun god Shamash receives homage in his shrine; his emblem, a solar disk, rests on an altar.

Zigzag beams radiate from a beaten-gold mask of Inti, the supreme god of the Inca people of Peru dating from AD 1200.

Setting undergrowth ablaze both improved sightlines and drove animals out into the open, where waiting hunters could be ready with spears and bows and arrows, or with nets for smaller animals. More deadly still, fires were set to startle grazing herds and direct their stampede toward cliffs or steep banks. Falling by the hundreds to the bottom of deep ravines, the terrified animals were finished off by a hail of spears and arrows, or by rocks showered down on them from above. Such strategies yielded meat in hitherto undreamed-of quantities: At the foot of one cliff in France, for example, the remains of more than 100,000 wild horses have been found. Much of the meat was wasted, since the hunters ended up with more than they could use or conveniently store, and animals lying buried beneath piles of heavy carcasses were often simply left behind to rot.

Between 14,000 and 10,000 years ago, hunting on a near-industrial scale, using ambushes and fire, probably contributed to a wave of extinctions, particularly among large herbivores such as the mammoth, woolly rhinoceros, giant deer, and American horse, and also among the great carnivores such as the saber-toothed tiger that had preyed on them. The effect was most pronounced in the Americas where, within 1,000 years of the arrival of humans across the land bridge from Asia, dozens of varieties of large mammals died out. Changes in climate and vegetation also played

their part—in particular, the advance of woodland into the grassland habitats favored by the larger herbivores. On the other hand, when such drastic transformations had last swept the earth—during a great warming period that had commenced some 120,000 years earlier, and before the arrival of human hunters on the scene—there had been no comparable cost in animal species. The picture is further confused by the fact that those animals known to have been most commonly killed in large-scale drives, such as the reindeer in Eurasia and the bison in North America, survived in abundance into modern times; many of Africa's large species also came through largely unscathed, despite hunting by man. The real reasons for the extinctions remain in doubt, although it is clear that humans had become powerful enough to at least influence developments in the natural world.

For all the efficiency of the new weapons and tactics, however, the relationship between the hunters and the hunted encompassed the spiritual as well as the pragmatic. Most hunting expeditions were still closely run affairs in which the wits and cunning of humans and animals were evenly matched. To have any chance of success, hunters had to possess a sure understanding of their prey's natural behavior; this understanding extended to respect and empathy, and partook of the hunters' religious sensibilities concerning the other forms of life with which they shared the earth. These feelings were given tangible form in an immense outpouring of artistic activity: Many thousands of personal ornaments and other artifacts have been found, and these must represent a mere fraction of the output of the period. Most are small and portable, in keeping with the nomadic hunter-gatherer existence: fine beads and necklaces fashioned from colorful stones and seashells, and remarkable human and

The round emblem painted on this limestone tablet in Sweden in the fifth century AD represents the sun, whose life-giving powers were invoked by early Norse peoples at midwinter festivals.

A wooden shield probably made in the eighteenth century AD bears the glaring visage of Sūrya, a sun god worshiped in India from 1500 BC until the present day.

animal figures worked in stone and clay, bone and ivory. More permanent works survive in the form of paintings executed on the walls of caves in which the hunters sheltered. The most celebrated surviving examples are at Lascaux in France and Altamira in northern Spain: Here, vividly captured in paints prepared from charcoal and colored earth, bison, wild bulls, stags, and horses buck and charge across the cavern walls, the flat images given depth by the contours of the rock, from which they appear to grow organically. Around these leaping forms, mysterious symbols complete larger designs. The meaning and purpose of these paintings—whether they played a part in fertility rites, represented favored totemic animals revered as protectors, or perhaps were simply decorative—cannot be known, but the feelings of awe that inspired their execution are still apparent.

Far to the south, similar paintings were produced by hunter-gatherers in caves along South Africa's Drakensberg region, a craggy escarpment snaking through Natal, Lesotho, and into the remote eastern part of Cape Province. And here, in territory generally considered useless by later civilizations, the descendants of the original artists have survived. These people are the !Kung—the exclamation mark represents a click of the tongue, much used in their language. (They were formerly known as the Bushmen, an appellation that in South Africa acquired derogatory associations.) Their modern domain lies some distance to the north of the Drakensberg, in the western part of the Kalahari Desert—a parched, sunbaked expanse of scrub and waving grass punctuated by occasional trees, with little in the way of surface water, covering thousands of square miles of southern Africa's interior. No respecters of modern political boundaries, the !Kung bands roam a large area straddling the border between Botswana and Namibia.

Although the Kalahari is by no means completely arid, its rainfall is erratic, and much of a hunter-gatherer band's energy is devoted to finding water. Longstanding, ritualized habits govern the use of this precious resource: The more obvious and lasting sources of water are left until seasonal, less easily exploited supplies are exhausted. Hunters who are traveling far from their camp, or from the low-lying land likely to have water available on the surface, carry water with them in the empty shells of ostrich eggs. When this supply runs out, a hunter is able to use a straw to reach water trapped in hollow trees or root systems, or suck water from beneath the sand through a tubular stem.

The writer Laurens van der Post, who was born on the edge of the Kalahari in 1906, once witnessed a hunter named Bauxhau extract water in this manner. The hunter first dug a hole in the sand about an arm's length deep. He then took the stem of a bush with a hollow core and wound some grass around one end, to act as a filter. He put the stem into the hole and packed sand tightly around it; then he arranged some empty ostrich-egg shells by its side. Finally, he took a short stick and placed one end in his mouth and positioned the other over one of the shells. When all these preparations were complete, he put his lips to the tube.

> *For about two minutes he sucked mightily without any result. . . . His broad shoulders heaved with the immense effort and sweat began to run like water down his back. But at last the miracle happened and so suddenly that I had an impulse loudly to cheer. A bubble of pure bright water came out of the corner of Bauxhau's mouth, clung to the little stick, and ran straight down its side into the shell without spilling a drop!*

When all the shells had been filled, van der Post himself attempted to repeat what the hunter had done, but could extract no moisture at all.

The !Kung's intimate knowledge of their environment is as essential to their hunting and gathering of food as it is to their finding of water. They take their sustenance from around 100 different animal species and 150 plants. Of particular importance is the mongongo nut, the fruit of trees that grow in groves in the sandy soil around rock outcrops. These highly nutritious nuts, which are protected from rotting after they have fallen to the ground by their hard shells, yield almost half of the calories consumed by the !Kung in their vegetable diet. The food the !Kung prize most, however, is the meat of large antelope such as the kudu, gemsbok, and eland, but usually they make do with smaller antelope and warthogs, and also with birds—brought down with arrows or trapped—and their eggs. Some species of lizard and snake are also eaten. Insects—apart from a few special delicacies—are not much valued, but the honey of wild bees is a very popular treat. One particular type of beetle performs an important function: It yields a venomous fluid that is fatal when taken in through the bloodstream but harmless when ingested orally. Applied with care to arrowheads, it turns small darts into lethal weapons, while leaving the resulting meat perfectly safe for human consumption.

In general, plant foods, abundant year-round in the warm climate of the Kalahari, form the staple of the !Kung diet, and animal protein is less important than it was for the early European hunter-gatherers. Although the !Kung do what they can to nurture grazing animals, burning large areas each winter to encourage the growth of new grass, game in the modern Kalahari is not as plentiful as it once was in Europe, and hunting is more difficult and time-consuming. Each animal kill, it has been estimated, requires an average of four days' hunting. Even after being hit by a poison arrow, a large antelope may still have to be tracked for several days before it can be finished off. Perhaps in response to the relative scarcity of big game, the !Kung have evolved a taboo on killing more animals than is necessary. They also tend to pursue surplus males among the animal population.

In most regions of the world, the hunter-gatherer way of life gradually yielded to other forms of activity more appropriate to settled and larger communities. But it was the norm for more than 90 percent of humanity's time on earth, and the system gave those who practiced it—as it still gives the !Kung—a simple yet adequate standard of living in return for a relatively small investment of effort. In the case of the !Kung, it has been estimated that a leisurely 2.4 working days per week are enough to provide for their needs. In addition, the nomads' extensive knowledge of the resources avail-

able in their territory, their physical mobility, and their dietary flexibility in responding to local gluts and shortages have enabled them to ride out natural disasters and climatic fluctuations more easily than their settled, specialized successors.

By 10,000 years ago, there were some 10 million people in the world. Small as this total figure seems by later standards, it represented, given the extensive territories required for the hunter-gatherer existence, the beginning of a crisis in population density. More intensive ways of reaping the earth's resources were required. The deliberate cultivation of crops and husbandry of animals would meet this need—yet it would lack the light touch of the hunter-gatherer ways. In making the earth work much harder, agriculture was to have a radical and enduring impact on it.

Hunter-gatherers themselves gradually began to control the growth of favored food plants, taking seeds from natural patches of vegetation and planting them in more convenient and more productive sites. Over the centuries, human intervention increased, with dramatic consequences. Plants taken under cultivation, harvested, and then resown year after year tend to change genetically in response. The main seed-bearing core of a wild plant is usually brittle, encouraging propagation by allowing the ripened seeds to scatter at the slightest breeze. The specimens favored by humans were those whose cores were tougher than average and, therefore, more likely to retain their seeds long enough to be reaped by the farmer's sickle. Through several generations, the resilient core became a standard, genetically ordained feature of the plant in its cultivated form—so much so that the plant became wholly domesticated, unable to release its own seeds, and dependent instead on its cultivators' stripping the grain from the stalk and planting it in the ground. In the same manner, when the farmers came to select the seed for each year's sowing, they chose the largest, plumpest grains; over time, in consequence, the average size increased, and the domesticated plant grew into a distinct form, often bearing little apparent resemblance to its wild ancestor.

Once plants had been taken under human control and were producing higher yields, some of the surplus generated could be used for feeding animals, thus accelerating their own domestication. Again, through many generations of human husbandry, domesticated animals grew into distinct forms, so genetically altered that they could no longer reproduce with their wild relatives. Unlike plants, animals usually became smaller in their domesticated forms. Size differences between males and females were eroded, and adult animals tended to retain juvenile characteristics of bone structure and coloring—piebald patterning, for instance, common in do-

The raging forest fire shown in this artist's drawing represents the most powerful impact of hunter-gatherers on their landscape. In addition to using fire to cook food and harden the tips of their weapons, hunters learned to set undergrowth ablaze to drive animals into the open or stampede herds toward a prepared ambush; the fires also encouraged the regrowth of tender young plants that attracted grazing animals. Over time, the recurrent effects of very high ground temperatures in forest fires caused the local extinction of certain plants while favoring the growth of others that could regenerate from under the ground or that had thick protective barks.

The flare of ripening wheat against the waters of the Euphrates River in present-day Iraq proclaims the human ascendancy over nature made possible by agriculture. Domesticated cereal crops, first flourishing in the Fertile Crescent in the eighth millennium BC, were soon seeded along the banks of the Nile River in Egypt, near the Indus River in Pakistan, and in the valleys of Central America. The labor of the humans the crops supported was changed from hunting and gathering to planting and harvesting, as shown on the right in an Egyptian tomb painting dating from 1300 BC. Overleaf, an Egyptian limestone statuette represents a maidservant grinding wheat into flour: Bowed to new rhythms, the human body learned new aches.

mestic animals and young wild ones, is unusual in adult wild animals. Domesticated animals provided not only meat, hide, wool, and hair but also dung for fertilizing the soil and labor for drawing plows.

In a range of environments across the world, a balanced selection of both plants and animals was brought under human management. In the Fertile Crescent in the Near East, early forms of cereals such as wheat, barley, and rye, and legumes such as lentils and peas were farmed alongside animals including cattle, sheep, and goats. Irrigation allowed the gradual outward spread of agriculture into the drier steppe areas of the Eurasian interior. In China and Southeast Asia, rice and millet provided the staple grains, cultivated with the assistance of domesticated water buffalo. Across the Pacific Ocean, farmers in South and Central America grew root crops, which included potatoes, yams, and cassava. In the high Andes, llamas and alpacas, carrying heavy loads up and down vertiginous mountain tracks, helped to make large-scale cultivation practicable.

Farther north, through much of Central America from the lush rain forests of the coastal plains to the sunbaked hill country of the interior, a different system of agriculture was slowly taking form. Some 7,000 years earlier, in the hot, dry Oaxaca Valley of southern Mexico, peoples who had lived for countless generations by hunting white-tailed deer and cottontail rabbits on grassy hillsides, catching mud turtles in ponds and creeks, and gathering a wide range of vegetable foods began to cultivate some of their most important plants as a standby to supplement what they could gather from the wild. Their squash, beans, and teosinte—a seed-bearing local grass—began to evolve into distinct strains, steadily increasing in size. Its seedpod growing from less than two inches in length to about eight inches, teosinte became the domesticated cereal known in modern times as corn.

For many centuries, crop growing in the Oaxaca Valley remained a supplementary activity, and it was not until some 3,800 years later that the evolving corncob grew to a size that allowed the Oaxacans to make agriculture their main pursuit. This relatively slow rate of development was to a great extent a result of the absence of large animals suitable for domestication. The addition of animal protein, manure, and haulage to the Oaxacan economy would have allowed a wholehearted commitment to the sedentary farming life much earlier.

Centuries later, North America's Iroquois Indians, themselves agriculturalists working in the tradition first established by the early farmers of Oaxaca, described the beans, squash, and corn they grew as "three sisters, those on whom our life depends." And although corn was certainly the staple crop in Central America, the other plants were no less vital. High as corn is in starch, it is relatively low in total protein; while it contains essential niacin, this occurs naturally in its bound form niacytin (early farmers learned to add a pinch of wood ash to their corn dishes to release the niacin). More important, corn lacks completely certain amino acids that are vital to human life. These amino acids are found in beans and squash, along with the extra protein that is needed for a balanced diet.

Corn, beans, and squash provided the first Oaxacan farmers with most of the nutrients they needed, and it was as a sisterhood that they were cultivated. Prodding the earth at regular intervals with wooden digging sticks in order to make holes, the Oaxacans planted the seeds of the three plants at different times, so that all would come to fruition simultaneously. The workers then toiled each day in the blazing sun, tending the seedlings individually to coax each into abundant growth. The beanstalks were carefully trained up the sturdy stems of the corn for support, with the squash hugging the ground in between. Since this system packed the plants into tight rows, leaving very little space in between, it was particularly important that any competition from encroaching weeds for precious mineral nutrients and light should be fiercely resisted; so the Oaxacans plied their wooden hoes busily throughout the growing process to keep the weeds at bay.

While the clearing of land, tilling of soil, sowing, constant weeding, and harvesting involved in agriculture of this sort demanded an enormous investment of labor, the return in produce more than made up for the effort. Central American communities, like other emerging agricultural societies throughout the world, found themselves able to support increasingly large populations, including many persons not directly involved in cultivation. As sedentary communities, moreover, they began to accumulate possessions—not only reserves of food but objects for

which such reserves could be traded. These acquisitions began with valuable tools and ornaments in numbers impracticable in nomadic times; eventually, the more successful farmers came to own parcels of land, buildings, and precious metals. Differences in wealth between individuals and families began to emerge, and with them came the evolution of landowning and laboring classes.

Small villages grew into great cities as the prosperity agriculture brought generated more work in nonagricultural occupations. Artisans were engaged in manufacturing farming tools and weapons, and in creating ornamental artifacts for the rich. Engineers and builders were employed to construct palaces and granaries, and staffs were hired to run them. As a means of consolidating their power, rulers gathered bodyguards and armies about themselves as protection against resentful inferiors and rival magnates; they also established bureaucracies to manage the institutions of state that ensured they would remain in power.

Powerful and complex as such new societies became, however, they were vulnerable in a way that hunter-gatherer groups—with their low population densities and generalist survival strategies, responding instantly and easily to the effects of local environmental changes—never had been. The greatest of the Central American empires, no less than the humblest village, was built on corn, beans, and squash—and if drought, flood, or any other natural disaster should cause the crops to fail, the edifice would crumble. This dependence was recognized: According to Spanish priest Bernardino de Sahagún, writing during the sixteenth century in the early decades of Spain's colonization of Mexico, an Aztec of the day described corn in religious terms. It was "our flesh, our bones," he said. "I honor it. I desire it. I venerate it, esteem it. It is our sustenance."

Similar attitudes prevail among the Hopi Indians of arid northeast Arizona, many of whom still wrest a living from the rocky, unyielding soils around their pueblos, cultivating corn and other crops according to the methods developed thousands of years ago by the first American farmers. For the Hopi, nothing can be more important than the successful outcome of the corn crop, without which the life of the community simply could not continue. The corn plant itself—with its upright posture and its torsolike cob, with pollen-bearing tassels hanging down like tresses of hair—is seen as resembling the human form. Each stage of the growing process—sowing, sprouting, maturing, and harvesting—is attended by ancient prayers and rituals, by communal dances and songs, all intended to encourage the corn as it struggles to grow. Each August, the men of the pueblo perform a snake dance, holding live rattlesnakes in their mouths, hoping that the reptiles will carry to the rain gods their pleas for showers to nurture the thirsty roots of the growing crops.

Just as Hopi rituals attend the annual progress of the corn crop, so the corn itself plays a leading role in all the myriad rituals surrounding the life cycles of men and women. When a baby is born, for instance, a "corn mother"—a cob with four kernels grouped in a cluster at its head—is placed beside the infant for its first twenty days, during which time the baby is kept in a darkened room, its body deemed to be still under the protection of its universal parents, the earth and sun. At regular intervals throughout this period, the baby is washed and then rubbed with cornmeal. A line painted on the walls and ceiling of the room with cornmeal completes the child's protection: Such lines are seen as barriers to evil spirits and are also used to block roads and trails. Each new day is saluted with cornmeal, scattered before the rising sun. Used at every religious ceremony, cornmeal's effectiveness as an offering stems from its traditional identification with the human body: As the Hopi sprinkle its seeds during prayer, symbolically they are offering up a little of their own flesh.

The Hopi have woven such a powerful web of emotion and religious devotion around their main source of sustenance precisely because they are uncomfortably aware, as were their predecessors, that this source may one day fail them, leaving them with nothing. Though relatively infrequent, disastrous crop failures have devastating consequences for farming communities and are remembered for many generations afterward. And if famine has proved an inseparable companion to agriculture since the first farmers tilled the soil, this has been in large part because the demands made on the earth by agriculture have at times simply been greater than the earth could bear. Cultivated year after year, fields become exhausted, drained by the hungry crops of all the nutrients and essential minerals they contain. Such land needs to be well rested. Farming communities discovered this early on in agriculture's history, but the unwary could still be brought to grief.

Even when apparently successful, agriculture took its toll on the earth. The clearing of forests for cultivation removed the extensive root systems that had locked soil firmly in place; once the trees were gone, the topsoil might easily be washed away by the first heavy rainfall. Trees are also far more efficient than other forms of vegetation in helping to dissipate moisture from the earth by evaporation. When they are cleared, the ground beneath loses this means of releasing its moisture, and apparently dry woodland might turn out, after deforestation, to be too damp and spongy for cultivation. It was this process that led to the development of upland peat bogs in some parts of Europe, such as the Dartmoor region of southern Britain.

The Mesopotamian goddess Inanna, shown above between symbols of fertility on a stone bowl made about 2500 BC, was one of a series of goddesses around whom legends were woven recounting the annual cycle of barren and regenerative seasons. Having descended to the underworld to seek her dead husband, Inanna was stripped of her powers but was later resurrected after being sprinkled with the water of life. The Egyptian goddess Isis, shown suckling her child in a faience statue from 1400 BC *(opposite)*, was married to Osiris, god of fertility and death; when Osiris was killed by the god of drought and his body scattered in pieces across the land, Isis restored him to life, and they conceived a child, Horus. In Greek legend, the mourning of the Greek goddess Demeter for her daughter Persephone, who was carried off to the underworld, caused the crops to die; the two were reunited by the intervention of Zeus, but Persephone was required to return to the underworld for a part of each year.

Another problem encountered by early farmers was salinization. In the Near East in particular, where river water was used to irrigate drier areas through systems of canals and ditches, the salt and other minerals carried along in the water, and normally destined to accumulate in the sea, built up in the soil instead. When the water evaporated, it left its salt deposits behind, to build up year after year. This process of salinization significantly reduced the fertility of the soil; once salt levels of more than one percent had been reached, the land became barren.

Other problems stemmed from the sedentary existence that agriculture dictated. Settled communities, packed together in villages or cities, in close proximity to their livestock and indeed to their own waste products, ran a risk of disease never imagined by earlier hunter-gatherers, who were protected by their low population density and their nomadic ways. Great pestilences periodically swept the world's agricultural areas, bringing horrendous loss of life. Always a very demanding system of land use in both labor and environmental cost, agriculture could at times prove disastrous to the well-being of its practitioners.

Some 4,000 years ago, on the central Asian steppe between the Caspian Sea and the northern edges of China, peoples who had for generations been reaping sparse harvests from arid and unrewarding soils began to cut their losses. They abandoned the agricultural life altogether. Since their oxen, their horses, and their flocks of sheep and goats had thrived on the limitless pasture of the steppe, it was naturally to this resource that they now turned.

The new pastoralists could not simply abandon their fields and continue grazing their old pastures, however. Because everything now depended on their flocks, they had to increase the numbers as well as the size of the individual animals in order to enhance their productivity as a food resource; this meant finding the best possible year-round grazing and watering. Localized grazing—seasonally fluctuating and quickly exhausted by large flocks—could never meet the new standards. Thus, pastoralists had to move their flocks from place to place, wherever the best pastures were to be found. Thousands of years after their ancestors had abandoned the wandering hunter-gatherer life, the people of the steppes found themselves nomads again.

If economic considerations made the new nomadism desirable, a number of other factors made it more practicable than it might have been in earlier times. Perhaps most vital of all was increased mobility. Over the centuries since its first domestication, the horse had been tamed to a still higher degree and was now thoroughly broken to the pack and the saddle, dramatically extending the range of human travel across the endless steppe. Travel was made even easier by the advent of the wheel, introduced from the agricultural Near East. While still having to shed many of the possessions that had surrounded them in their sedentary days, the new nomads could use carts hauled by oxen or horses to carry loads never dreamed of by the hunter-gatherer bands of earlier ages.

Within the human body itself, meanwhile, a more intimate, physiological development had slowly and imperceptibly been taking place. The enzymes secreted in babies' stomachs to permit the digestion of their mothers' milk, previously lost after infancy, were beginning to endure throughout adulthood. The impact of this change was revolutionary, for it added a second, well-nigh-inexhaustible source of rich animal protein to the pastoralists' diet. Where animals had once provided nourishment solely in the form of meat—and were productive only when killed—goats, sheep, and mares were now twofold nutritional resources, supplying not only meat but also more long-term nourishment in the form of daily milk. Although milk was often drunk in its natural form, a range of dairy products were developed to make it more readily storable and portable: yogurt, cheese, and butter. In addition, dehydrated milk was prepared by heating the liquid until the milk separated and drying the resulting solids in the sun.

Mare's milk was especially prized by the pastoralist bands. Oxen were kept in small numbers for haulage, but except in Africa, cattle generally proved ill-adapted to the nomadic lifestyle because they consumed large amounts of water and high-quality grass in return for milk that was thinner and less rich in vitamins than that of the mare. Mare's milk could also be fermented to make an alcoholic drink, which later Mongol nomads would call kumiss.

Milking was not the only means the pastoralists had of drawing nourishment from living animals: Often it was tapped more directly in the form of blood, which could be used as an ingredient in cooked food or simply drunk neat. As long as only small amounts of blood were taken at a time, and the animals were bled in rotation to allow each a few days to recover before its blood

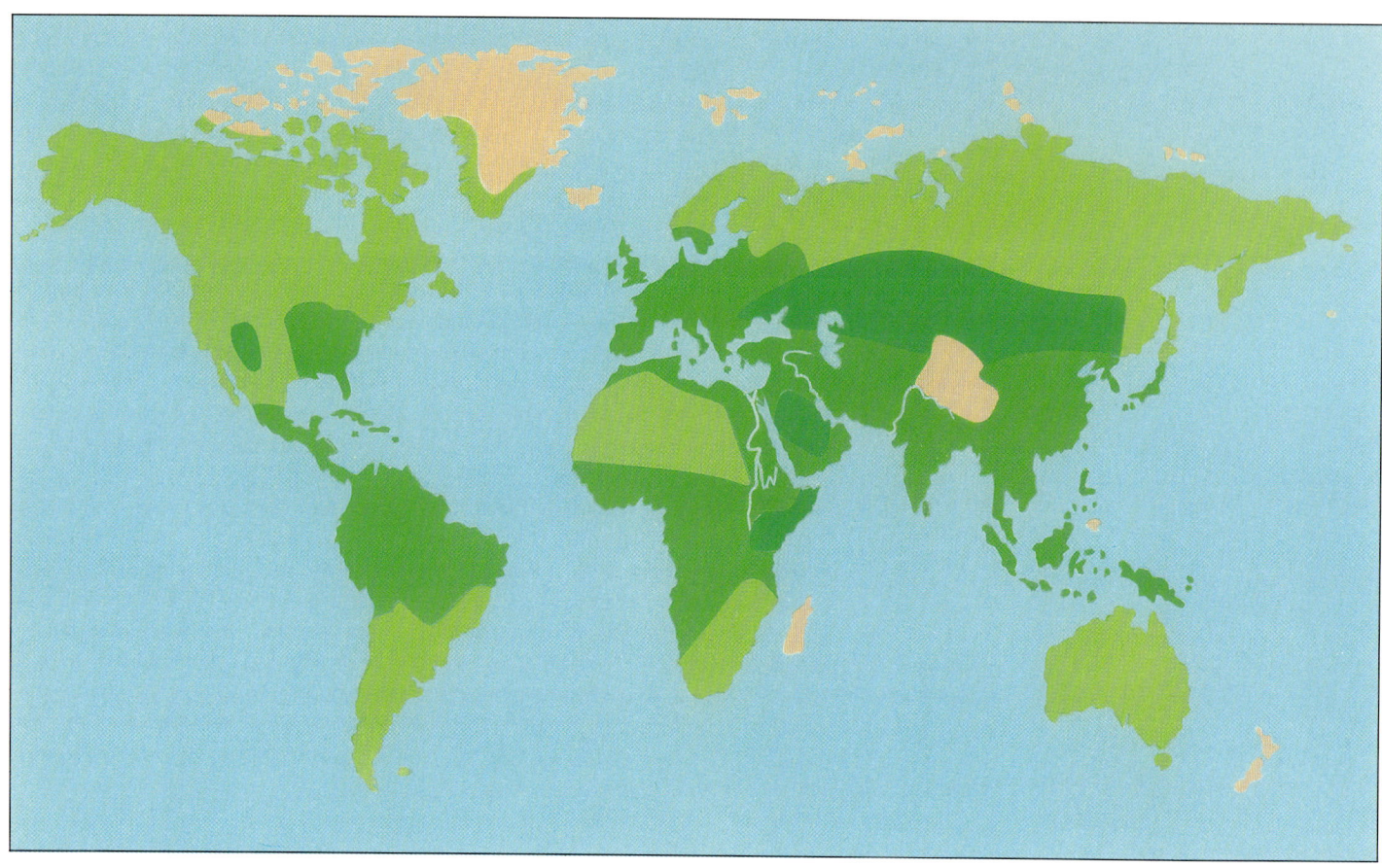

This map shows the approximate regions of the world occupied by hunter-gatherers *(light green)*, by agriculturalists *(medium green)*, and by pastoralist herders *(dark green)* very early in the first century. The areas colored beige were as yet unclaimed by humans. Relative to the timespan of human inhabitation of the earth, the encroachment by farmers and pastoralists over fewer than 8,000 years on territory first occupied by hunter-gatherers was spectacularly rapid. This expropriation of land by late-developing cultures was to be continued. City dwellers who depend on others for their food now constitute more than 40 percent of the human population, while hunter-gatherers have dwindled to less than 0.001 percent.

was taken again, the practice could continue indefinitely without any adverse effects on the pastoralists' livestock.

The pastoralists now had in their flocks and herds a food supply adequate for all their nutritional needs. As long as their livestock could be fed, they themselves did not go hungry. So, traveling for much of the year in small family groups—for larger social gatherings would put overwhelming pressure on an area's pasture—the nomads would make their way across the steppe, moving on at regular intervals to new, unexploited pastures in endless pursuit of good grazing. Their unhurried flocks feeding as they went, they covered no more than a few miles a day, but there was little need for haste. There was usually a destination, however, and a timetable of sorts. While a few groups wandered without definite itinerary, permanently on the move, more often the nomadic pastoralists followed well-tried routes, moving each year between winter and summer grazing areas often many hundreds of miles apart. Sometimes, the migration would be from an area of low ground—mild and grassy in winter, but hot and sparsely pastured in summer—to an upland region that remained relatively cool, moist, and green in summer, although it might be impossibly harsh in winter. In other cases, the move would be a straightforward journey from north to south, taking advantage of changing climatic zones.

The flocks of the nomads provided far more than a source of food. Their hides were used for making harnesses, their bones for tools. Their dung provided the perfect fuel far out on the steppe where wood was very scarce. Wool was spun into cloth for garments and for the elaborately woven rugs that were used as portable flooring in the nomads' tents. Wool was also used to make felt for the tents themselves. After being spread out in great quantities on the ground, the wool was painstakingly flattened and rolled into tight cylindrical shapes until the matted fibers cohered as a single sheet. These sheets, waterproofed with animal grease, and in winter often several layers deep, were then stretched on frameworks of wooden poles and held down by weighted horsehair ropes.

Though in principle self-sufficient, the nomads' existence was often insecure. The lands they had abandoned as farmers had been marginal, but later agriculturalists often succeeded in bringing them under cultivation by using systematic irrigation, and the nomads found themselves pushed deeper into the barren interior of the steppe. Even good pasture was subject to occasional drought, while a single late frost, destroying the young shoots of spring, could blight a season's grass over vast areas.

By the time this ivory plaque of a cow suckling its calf was carved in Syria in the eighth century BC, milk was sustaining humans as well as the offspring of wild animals. Both hunters and farmers contributed to the domestication of animals. The hunters brought home the young of the prey that had been killed, and the farmers kept the captured animals for eating at a later time, fattening them on surplus crops. Together, the members of the band gradually exerted an increasing degree of control over the migrating herds that they followed. By 4000 BC, the basic herd animals—sheep, goats, cattle, and pigs—were serving humans in ways unknown to earlier hunter-gatherers: They provided milk, wool, and power for drawing plows, and their possession was a new indicator of wealth.

Like the agriculturalists, moreover, the nomads contributed unwittingly to their own problems, their very success sometimes leading to setbacks. If their flocks grew too large, they placed a heavy burden on the available pasture. Persistent overgrazing could leave the lushest grassland bare and destroy what trees and shrubs there were on the steppe. Without such cover, nothing kept the topsoil from being carried away by wind and rain, after which the land became useless desert.

The nomads seldom enjoyed the sort of surpluses the agriculturalists did, and often suffered severe shortfalls. They were frequently anxious to trade livestock, felt, leather goods, and woolen carpets and clothing for the produce of the settled agriculturalists near their migration routes—but the farmers, who generally regarded their wandering cousins as irresponsible and untrustworthy vagrants, were often less eager to do business. The nomads therefore sometimes resorted to war to get what they required, making mounted attacks on village settlements to loot the contents of granaries and storehouses and to seize the villagers' livestock. Such were the nomads' easy mobility, their consummate horsemanship, and their skill in shooting their short bows from the backs of their galloping horses that they were usually victorious.

Periodically, nomad bands would push outward from the interior of central Asia, massing in loose confederacies to seek both grazing and conquest. Their descendants would bring down empires in China, and in western Europe, they would destroy the might of classical Rome. But usually the nomad alliances broke down quickly once victory had been won—their social order was by definition a shifting, temporary thing. Outnumbered by their subjects, some conquering nomads adopted settled ways and were absorbed into agricultural communities; others quickly quarreled and parted, to range the steppe once more.

In the Zagros Mountains of southern Iran, to the south of the Caspian Sea, a people known as the Qashqāī have kept the nomadic ways alive into modern times. Although cultivated fields and the factories and built-up areas of the city of Shīrāz now block their ancestral migration routes, the Qashqāī still manage to make the round trip between their summer and winter quarters each year, some clans covering 600 miles with their flocks of sheep and goats. Members of the tribe's hereditary nobility now own tracts of agricultural land in each area—apple orchards and vegetable fields in the north, citrus and date orchards and fields of barley and wheat in the south—and travel between them by truck and Jeep. But most ordinary Qashqāī still depend on their flocks for sustenance and make the annual migration in the traditional way, driving their flocks across rugged terrain on horseback and on foot, and living all year round in tents made from woven goat hair.

Apart from an assortment of horses, donkeys, and camels, the average Qashqāī family owns a flock of around forty sheep and thirty goats, from which it gets meat, milk, and dairy products, as well as wool and hair for making into cloth. These animals are common to most nomadic pastoralist peoples; camels are generally herded only in the driest regions, and cattle only where there is plentiful rain and a good supply of grass. The goats kept by the nomads provide a vital safety net: These hardy creatures are able to survive in conditions that would kill off most other animals, and they are the last to cease giving milk when the land is stricken by drought. All animals kept by pastoralists tend to be lean and spare, but nevertheless constitute their owners' chief form of wealth. (The word *pecuniary*, meaning to do with money, derives from an ancient Sanskrit word for cattle through the old Latin *pecu*, meaning cattle as wealth or property.)

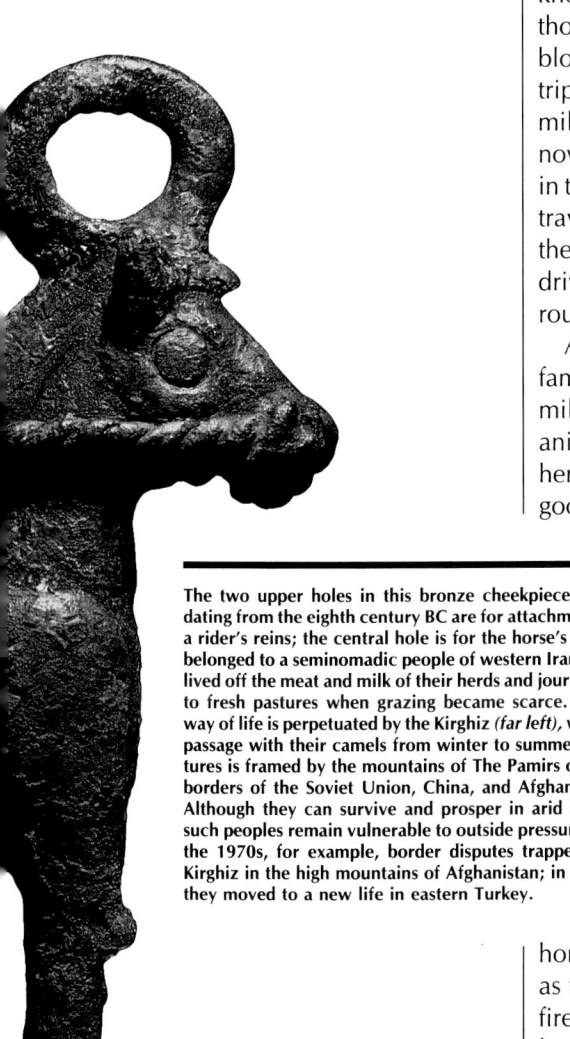

The two upper holes in this bronze cheekpiece *(left)* dating from the eighth century BC are for attachment to a rider's reins; the central hole is for the horse's bit. It belonged to a seminomadic people of western Iran who lived off the meat and milk of their herds and journeyed to fresh pastures when grazing became scarce. Their way of life is perpetuated by the Kirghiz *(far left),* whose passage with their camels from winter to summer pastures is framed by the mountains of The Pamirs on the borders of the Soviet Union, China, and Afghanistan. Although they can survive and prosper in arid lands, such peoples remain vulnerable to outside pressures. In the 1970s, for example, border disputes trapped the Kirghiz in the high mountains of Afghanistan; in 1982, they moved to a new life in eastern Turkey.

The Qashqāī winter in the green hills of Khonj above the Persian Gulf, where the men tend their flocks and hunt ibexes and other game for sport—like their central Asian forebears, they still shoot from horseback with deadly accuracy. The women look after their families, spin wool, and weave carpets. Each year in late March, the community dismantles its tents, packs up its belongings, and loads everything onto the backs of horses, camels, and donkeys to decamp as one, driving its flocks northward, raising as they go a dense pall of thick dust that seems like the advancing smoke of a great fire. At the end of each day's journey, the men raise the roofs of their family tents on long poles and fasten the sides with wooden pins, while the women and children pound stakes into the hard earth. Rugs are shaken and spread, pack animals are watered and fed, and fires for cooking are lit.

A present-day member of the Qashqāī remembers in his youth seeing a press of 5,000 families crowding through mountain passes and crossing fords, accompanied by perhaps 500,000 sheep and goats and some 70,000 horses and donkeys. Their numbers now are much diminished. Many of their traditional pastures were nationalized in the 1950s and the land was redistributed to settled farmers; new roads were cut across their grazing lands, destroying the grass; and in tent schools erected in their summer and winter camps, the Qashqāī children were taught to read and write, enabling them to get jobs in towns. But for those who have not forsaken the old ways, the rhythm of seasonal migration continues, and for almost two months, the Qashqāī continue their journey, piloting their animals across apparently trackless wastes, steering them through fast-flowing rivers, and negotiating treacherous crags and rockfaces, until they arrive at their summer encampment. A few short months later, as autumn arrives, it is time to begin the return trip once again.

The Qashqāī of Iran, the Hopi of Arizona, and the !Kung of the Kalahari Desert are rare survivors of an ancient lifestyle. In most parts of the world, the techniques of survival of the first human inhabitants—which evolved primarily in response to local conditions—have long since been replaced by strategies based on the deliberate adaptation of nature to meet human needs.

Over the last five centuries, many hunter-gatherer peoples in particular have been driven to extinction by the weaponry of invading armies or by the expropriation of their foraging grounds; those who escaped this fate live mostly in desert, rain forest, or Arctic regions, where agriculture is impracticable or uneconomic, or have been reduced to dependency on reservations. They have suffered too from the arrogant misconceptions of succeeding cultures, epitomized by the opinion expressed by the English political philosopher Thomas Hobbes in 1651: "No arts, no letters, no society, and, which is worst of all, continual fear and danger of violent death, and the life of man solitary, poor, nasty, brutish, and short." Only one of these generalizations—the absence of letters, or a system of writing—accords with reality.

Because of the endurance of such prejudices, the fact that the !Kung and peoples like them should choose to continue with their ancient ways seems surprising to modern city dwellers. Doubtless, the life of the cities is as perplexing to them. Arduous and materially impoverished as their lifestyles may appear, they endure precisely because they have succeeded in meeting the needs of their practitioners over many centuries, and because, despite disruption by modern cultures and governments, they continue to meet those needs. Their persistence represents not a failure of development but the triumph of economic systems so perfectly matched to their environments that no further development was required.

SANCTUARIES OF THE SPIRIT

The earliest humans, watching lightning flare in the sky or storm-driven waves thunder against the face of a sheer cliff, could not doubt that the earth was spirit as well as matter. And the spirit was manifest not only in the awesome spectacles of weather but also in the daily round of humankind's existence: At times, the earth provided abundantly for all human needs; at other times, it withheld its bounty. Clearly, the natural world was infused with energies beyond human ken.

In all parts of the world, certain places in which the immanence of the earth's spirit was most acutely felt were accorded special reverence. Prominent among these sacred sites were particular mountains, lakes, rivers, waterfalls, caves, and groves of trees. Many of these sacrosanct places were distinguished by the extreme beauty of their natural formation or setting, as shown on the following pages. From the tumbling waters of the 740-foot-high Kaieteur Falls in Guyana *(right)*, for instance, spray dances in clouds of ever-changing hues. When drought reduces the volume of water, a stone shape can be seen at the base that is believed by local people to be the petrified canoe of a king who offered himself to the spirit of the falls.

The web of legends and myths that accumulated around all such sites gave narrative form to their inherent natural forces. Rites of worship were invoked and religious cults were formed to protect holy places from desecration, and severe penalties were exacted for their defilement. In the first century AD, for example, a Roman general who cut timber for shipbuilding from a grove dedicated to the god Asklepios on the island of Kos was executed in the grove itself.

Wherever the bounds of the sacred were transgressed—whether by an army of occupation or by a company hoping for financial gain from mineral deposits—the pain of violation was felt by all who deemed the site holy.

Dyed red by the setting sun, the furrowed sandstone cliffs of Ayers Rock rise above the barren plains of Australia's desert interior like the humped back of some stray subterranean creature. Five miles in circumference and more than 1,100 feet high, the rock was named after the premier of South Australia around 1873 by an explorer ignorant of its sacred status in the mythology of the Aborigines, to whom it was known as Uluru. The origins of the caves and potholes that puncture the rock's smooth surface are accounted for in the legends of the Aborigines concerning their ancestors, and Ayers Rock has now become a place of secular pilgrimage for thousands of tourists every year. In other parts of the world also, certain mountains and high places have remained sacred to successive cultures. Mount Ararat in eastern Turkey, for example, was associated in local Armenian lore with a great flood that engulfed the world prior to its identification in the book of Genesis as the place where Noah's Ark came to rest. In Japan, the serene, symmetrical Mount Fuji was held sacred by the aboriginal Ainu people centuries before a shrine was built on its summit by adherents of the Japanese Shinto religion.

Appearing as an artery of molten silver in the pale morning light, the Vilcanota River meanders beneficently through the Urubamba Valley in Peru, which was the heartland of the Inca empire from the twelfth to the fifteenth centuries. Temples built in the valley testify to the reverence accorded by the Incas to the river's life-giving waters that nourished the agriculture on which their prosperity was based. The Vilcanota was to the Incas as the Nile was to the ancient Egyptians, who 4,000 years ago addressed a hymn to the spirit of the river "who issues forth from the earth, who comes to bring life to the people of Egypt." In India, the Ganges River still attracts such worship: In 1986, four million pilgrims attended a festival at a town located in the foothills near the river. The Ganges for Hindus is Ganga, the goddess of purification. A verse from the Sanskrit epic *Mahabharata* declares: "If only the bone of a person should touch the water of the Ganges, that person shall dwell honored in heaven."

A circle of standing stones echoes the rounded shape of the mound behind it. This circle in Cumbria in northwestern England is one of many that were erected in northern Europe—more than 900 in Britain alone—by farming communities over a period of 2,500 years beginning in the fifth millennium BC. The avenues framed by 3,000 standing stones at Carnac in Brittany and the earliest megalithic structures at Stonehenge in southern England were also built during this period. The function of the circles was probably in part astronomical, the siting and arrangement of the stones being related to movements of the sun, moon, stars, and planets. But whatever their precise purpose may have been, the space they encompassed was hallowed ground. Through the pattern of the stones, humans sought to relate their existence to the more mysterious order of their natural environment: The erection of the stone circles was an act of homage.

THE MEDITERRANEAN WORLD

Around 700 years before the birth of Christ, two brothers in Boeotia, in what is now mainland Greece, had a disagreement over money. Both worked their own small farms, but while one was an industrious and prosperous farmer, the other, Perses, was a lazy and careless wastrel who eventually had to fall back on his brother for a loan. Perses got more than he bargained for. Rather than simply hand over money or grain, or lend him a plow, his brother decided to give him something far more valuable. He wrote for Perses an epic poem that contained in about 820 lines all the advice a farmer could need in order to prosper: the specific age of the oxen to buy for pulling a plow, the precise dates on which various crops should be planted and harvested, and the way to store surplus crops to fend off future hunger. Perses also learned how to avoid offending the gods upon whom much of his prosperity would depend. Urinating facing the sun, for example, was bound to upset Helios, the sun god. Armed with his brother's wisdom, Perses could not fail to do well for himself, although it would not be easy—the key to survival, the poem warned, was hard work.

Perses's brother Hesiod, the author of this hymn to prudent husbandry, was a farmer-poet who, it was said, learned the art of poetry from the Muses while he tended his sheep. It was also reported that after Hesiod was killed—as punishment for fathering an illegitimate child—his body was tossed into the sea but was borne back to shore by dolphins.

Much of Hesiod's poem, entitled *Works and Days*, portrays a landscape not unlike that of present-day Greece, where farming is a constant struggle against poor soil, barren slopes, and scorching, rainless summers. Never far from Hesiod's mind was the threat of hunger, even of starvation. And if the life of an individual was precarious, so too was the survival of whole early populations. For all the material wealth and cultural achievements of successive civilizations inhabiting the Mediterranean region, famine remained a constant specter. Valleys were filled with the golden wheat of scattered farms, and the foothills were covered with the silver green of olive trees and the darker green of grapevines, but that was not enough. Beyond the cultivated slopes rose white limestone mountains, some as much as 10,000 feet high, gouged with deep ravines and gorges. In places, these slopes were forested with pine and oak trees, but more often they were covered with maqui—a dense scrub of juniper—shrubby oak, and wild olive. A drab brown in the summer and a blaze of color in the spring, this terrain was virtually useless for agriculture throughout the year, good at best for rough grazing. Elsewhere, many streams ran dry in the summer, while in the winter, driving rain washed fertile soil from the hills and swollen rivers carried precious minerals out to sea.

The achievements of the peoples who inhabited the Mediterranean region derived above all from their delicate relationship with this obstinate and often inhospitable

This Greek terra-cotta tableau depicting a farmer trudging behind his ox-drawn plow dates from the seventh century BC, some 400 years after iron was first used in the Mediterranean lands to make plowshares. The eighth-century-BC poet Hesiod recommended that the plow beam and the stock—the part held by the farmer—be made of oak, and that the plow be yoked to two nine-year-old oxen. Where the soil was heavy, a team of as many as eight animals was required. Hesiod also suggested that the plow be guided by a forty-year-old laborer "who will keep his mind on work, not look around for friends, the way a young man would."

landscape. The natural world in which they lived shaped their development: It was the abundant fertility of the plains of Mesopotamia that allowed the Sumerians to build the world's first cities; without the mountains that restricted the size of Greek city-states to manageable proportions and without the sunlight that bathed their open-air meeting places, the Greeks might not have been able to develop the prototype structures of political democracy.

This shaping was not just a one-way process, for in their struggle to overcome the limitations of climate and landscape, the peoples of the Mediterranean exerted strong influences of their own. At first, when the population was small, farmers and pastoralists were able to satisfy their needs by living in cooperation with the seasonal rhythms of the land. But as their numbers grew and food supply became a pressing problem, they resorted to increasingly ingenious ways of working the soil and making the most profitable use of their limited resources. In doing so, they altered the face of their homeland and discovered that not only rivers and crops but also civilizations may be subject to rhythms of growth and decline.

About 10,000 years ago, when nomadic peoples began to assemble in settled communities in the Fertile Crescent around Mesopotamia for the common purposes of planting crops and breeding animals, the environment itself provided the solution to most of their practical needs. Every spring, the snows of the Zagros Mountains and the Armenian highland melted, swelling the Tigris and Euphrates rivers to coffee-colored torrents that washed over the flatlands. The floodwaters carried with them a mineral-rich alluvial sediment that was made up of fine soil particles mixed with clay and gravel. This material, known as silt, was the Mesopotamian farmer's primary source of fertilizer for the soil on which the richness of his crops depended. The floodwater also served to irrigate the land. In addition, the rivers provided a simple means of communication and transport, so that surplus produce could be exchanged with other communities.

Faced with the natural rhythms of flood, seedtime, and harvesttime, the Sumerians of Mesopotamia assumed they participated in a divine order, and that their society should be a reflection of the orderly ways of the gods. Their religion centered on the worship of an earth mother, and because they believed that the embodiment of male fertility was the bull, they marked the seasonal decline and renewal of the soil by ritual bull sacrifices. There were other early agricultural societies that followed similar fertility rites, some involving human sacrifice, usually of a king who had ruled for a year before being offered up to the goddess—the spilling of his blood represented the fertilization of the soil.

The farmers prospered, and by 3300 BC, they had established the world's first urban center at Sumer, on the plain between the Tigris and Euphrates rivers. Shortly afterward, the city of Memphis was built in Egypt on the banks of the Nile River. As would become the pattern throughout the Mediterranean, these cities were founded on agricultural prosperity: Only in those areas where the land provided enough food to sustain people not directly involved in its production were trade, specialized crafts, and an urban way of life able to develop. But while the new cities themselves accrued great wealth, especially for the emergent ruling classes who controlled the labor of the majority, they changed the whole relationship between their inhabitants and the land from which they had come. The fertile countryside was now under the political control of the cities and subject to new and exacting demands.

Steps cut into a hillside on the Greek island of Corfu permit the steep, rugged terrain to be put to maximum productive use. Throughout the Mediterranean region, such hills were once covered by forests; after the trees had been felled for fuel or construction, terraces such as these reduced soil erosion and retained moisture and, beginning around 3000 BC, were planted with olive trees. The Greek amphora *(inset)* dating from approximately 520 BC is decorated with a harvesting scene: The olives are knocked down with long poles, then gathered into a basket. The jar itself was probably used for storing olive oil, which was used in cooking, for lighting, and in the absence of soap, for cleaning the body.

A Sumerian poem known as the *Epic of Gilgamesh* recounts a legend that encapsulates this change. The hero Gilgamesh, ruler of the city of Uruk, determines to challenge the god who protects the forests: "I will conquer him in his cedar wood and show the strength of the sons of Uruk; all the world shall know of it. I am committed to this enterprise: to climb the mountain, to cut down the cedar, and leave behind me an enduring name." The Sumerian cities did indeed require enormous amounts of timber for house construction, pottery firing, and metal-ore smelting. Gilgamesh's successful completion of his task invoked the wrath of the sky god Enlil, however, who threatened a terrible revenge: "From henceforth may the fire be on your faces, may it eat the bread you eat, may it drink where you drink." The practical consequences of the clearing of the woodlands around the headwaters of the Tigris and Euphrates rivers were no less severe: Soil that had previously been anchored to the hillsides by the roots of the trees was now washed down by the rivers, gradually clogging irrigation canals and turning the lowland plains into a barren, salty waste. Deprived of its abundant agriculture, the Sumerian civilization soon weakened and lost its dominance.

Meanwhile, the people of Egypt's new cities continued to flourish along the narrow belt of fertile land on both sides of the Nile, not the least because they had found a source of timber that did not deplete their own limited supplies. Paintings in tombs at Thebes record the visits to Egypt of traders from the three-peaked, wooded island of Crete, which had become one of the Mediterranean's most powerful states by 2000 BC. In the Near East, where cedar was in short supply, its use was reserved primarily for the building of palaces and temples; on Crete, however, it was so abundant as to be used for everyday objects. In addition, the island was richly endowed with oak and pine.

Trade brought great wealth to the Minoan people of Crete, and their rulers built themselves huge palaces at Phaistos, Mallia, Zakro, and preeminently, at Knossos. In time, each palace became the hub of a local hierarchy—made up of bureaucrats, artisans, traders, priests and priestesses, and agricultural workers—that provided a degree of centralized authority over local agriculture and trade. At the same time, the religious practices of earlier cultures were echoed in Minoan rituals that centered on the bull: The god-king, known as Minos, was believed to be the human incarnation of the bull spirit, and on ceremonial occasions, he wore a bull mask during the performance of fertility rites.

In a Roman stone relief, two laborers pound rods to assist their stamping feet in crushing grapes to make the wine that the Romans, like the Greeks, often drank (diluted) in preference to water. The resultant mash was squeezed by a mechanical press, and the juice was poured into resin-coated vats to ferment. The fourth-century-BC plate on the opposite page is decorated with images of a red mullet, a bass, a torpedo fish, a sargus (now extinct), and a cuttlefish. In the time of Homer in the ninth century BC, fish was regarded as a food of the poor; in later centuries, when wild game was depleted and arable land was used for crops rather than stock raising, fish was for all classes a valued supplement to their grain and vegetable diet.

As the palace economies thrived, wood was overtaken in importance by other commodities. Hilly and sun-drenched, Crete was blessed with sufficient rainfall to nourish diverse forms of farming. Among these was sheep raising: The palace at Knossos alone controlled flocks that totaled more than 100,000 sheep. Although the island lacked the plains necessary for large-scale crop farming, fields and orchards yielded beans, lentils, peas, lettuce, and fruit such as figs and plums. The sea provided octopus, squid, and fish. Resin from the cypress pine was used for incense and as a varnish for preserving leather and metal. And as well as having storage facilities for surplus grain—which guaranteed food distribution in times of poor harvest—the palaces were stocked with giant ceramic jars containing wine and olive oil, Crete's most important exports.

Wild grapevines grew in abundance on the hilly island, and the Minoans soon discovered the pleasing qualities of fermented grape juice. Wine would become as common a beverage in Crete and Greece as beer had become in Mesopotamia, where a large proportion of the grain production had been consigned to fermentation.

Olives, too, grew wild on Crete, and in time they became even more important than grapes throughout the Mediterranean region—the Greek dramatist Sophocles would later describe the olive as the "sweet gray foster nurse" of his race. The olive tree is notoriously slow growing, taking at least ten years to reach maturity, and its cultivation required both forethought and the willingness to invest resources and labor in a product with only long-term gain. But the olive was also renowned at the time for its hardiness—it could survive droughts that killed off cereal crops—and for the ability of its thin, shallow roots to draw sustenance from barren soil in which little else would grow. As early as the end of the third millennium BC, the Minoans were cultivating the silvery-leaved trees in extensive olive groves. The oil extracted from the fruit proved to be one of the most versatile and valuable products of the Mediterranean world—a fuel for lamps, a basis for perfume or cosmetics, and, most important, a nutritious food that was more dependable than cereal crops.

But while the olive, once matured, renewed itself yearly, useful construction timber was altogether less inexhaustible. In order to provide building wood and fuel for potters and metalworkers, the palace at Knossos—which increased by twenty-eight times its original size within 1,000 years—required extensive forests to be cut down. Builders gradually began to use less timber in house construction, and potters sought alternative fuels. Clearly, there was not enough forest left on Crete to supply the Minoans' own needs, let alone those of their trading partners. Even the Minoan merchant fleet, lacking supplies of timber for repairing its vessels and building new ones, began to deteriorate.

By 1450 BC, the Mycenaeans of the Greek mainland had not only profited from the Minoans' decline but had also taken over the island of Crete itself, and they proceeded to establish a rich trading empire of their own around the Mediterranean. Timber again was the mainstay of their economy, especially in demand as a fuel for the Mycenaean bronze and pottery industries: The town of Messenia was home to at least 100 bronzesmiths, and pottery was produced on a scale for mass consumption and an international market that spread as far as Italy, Cyprus, and Palestine. But as agriculture boomed and the population rose, and as cities expanded in size and new settlements were created, land and timber became more scarce. By 1300 BC, woodcutters were being forced to venture into the heart of the Greek peninsula in search of suitable trees to fell.

When the cutters had finished their work, nothing protected the earth from the heat of the summer sun or the damage wrought by the driving rains of winter. Soil was washed away by streams that flooded, the slopes around them having lost their capacity to hold water. The plains flooded, too, and crops were smothered by the silt

In a relief inlaid with shell dating from 2400 BC (above), a ram is sacrificed in Mesopotamia, where sheep entrails were read by diviners who claimed to foretell the future. The thirteenth-century-BC Mycenaean pottery vessel on the right, cast in the shape of a bull's head, was used for offering libations: In the same way that animals were offered up to the gods, liquids such as wine or oil were poured through the bull's mouth onto the ground to propitiate the spirits of the earth or the underworld.

that was deposited by the torrents. Despite the construction of dikes and ditches to divert the water, many regions were by that time beyond reclamation—rich, brown forest soil had been eroded down to the level of red subsoil or white limestone bedrock. To add to the devastation, a period of drought was responsible for the failure of successive harvests. The result was an agricultural crisis compounded by a decline in the essential Mycenaean trades of metallurgy and pottery, which were dependent on now-scarce wood. Whole towns whose economies had been based on these industries were probably abandoned. The population figures plunged. In Arcadia, a mountainous inland area that still retained its forests, refugees from famine were able to survive only by eating acorns.

With the conclusion of the Mycenaean domination of Greece, the land itself fell into a period of decline. Sheep and goats that were pastured on marginal land in the foothills made their own depredations, destroying young shoots that bound the earth together and wearing away terracing that had been built previously for the cultivation of olive trees. In Greek Asia Minor, the effects were still apparent at a much later date: The Greek geographer Strabo noted in the first century BC that so much land was being washed away by one particular river that farmers had decided to sue it. "Lawsuits are brought against the Maeander River for altering the boundaries," Strabo reported, adding that when the river was convicted, "the fines are paid from

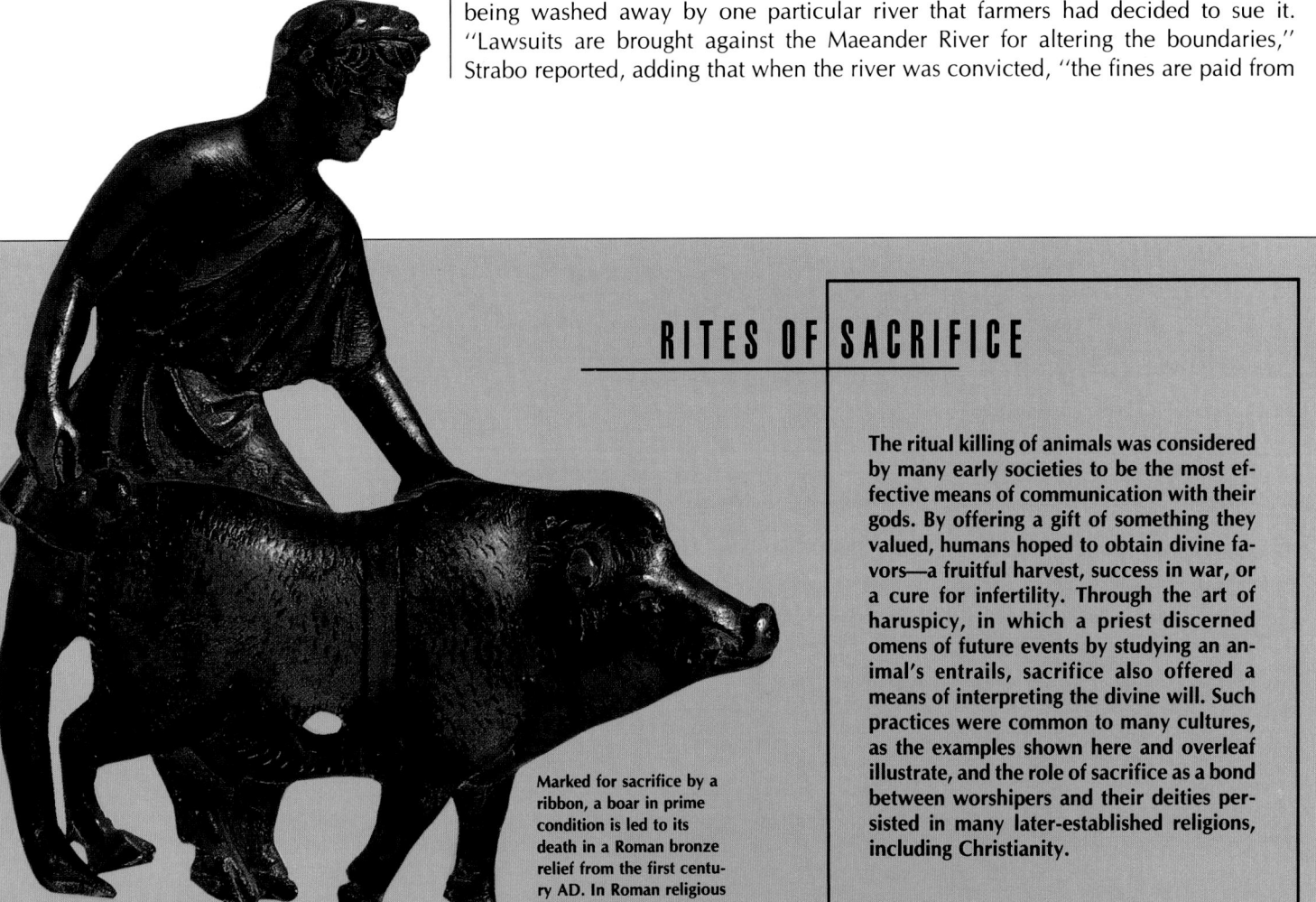

RITES OF SACRIFICE

The ritual killing of animals was considered by many early societies to be the most effective means of communication with their gods. By offering a gift of something they valued, humans hoped to obtain divine favors—a fruitful harvest, success in war, or a cure for infertility. Through the art of haruspicy, in which a priest discerned omens of future events by studying an animal's entrails, sacrifice also offered a means of interpreting the divine will. Such practices were common to many cultures, as the examples shown here and overleaf illustrate, and the role of sacrifice as a bond between worshipers and their deities persisted in many later-established religions, including Christianity.

Marked for sacrifice by a ribbon, a boar in prime condition is led to its death in a Roman bronze relief from the first century AD. In Roman religious belief, the pig was sacred to the goddess Demeter.

the tolls collected at the ferries." The bare-bones appearance that is now characteristic of some parts of Greece became so familiar that later generations assumed it was the natural and immemorial face of the land.

Other regions of the Mediterranean were not so diminished. The Phoenicians, from the area that came to be known as Lebanon, established an extensive maritime empire that was based on timber and textiles, the latter often dyed with secretions obtained by boiling vats of small sea mollusks. (The name *Phoenician* probably derives from a Greek word meaning "purple dye," referring to the expensive color that in many regions was reserved for the use of the highest members of society.) Around 700 BC, the Etruscans, from the western coast of Italy along the Tyrrhenian Sea, also joined the trading routes, capitalizing on their extensive reserves of metal ore: Their hills held deposits of copper, lead, tin, and—most important—iron, which had become the most sought-after metal for the manufacture of plow blades, tools, and weapons. And the people of Greece, too, gradually learned to survive and even prosper on their altered landscape, bowing their heads to the soil and often looking back with longing to a golden age when, in the words of Hesiod, "ungrudgingly, the fertile land gave up her fruit unasked."

Hesiod attributed the decline of the Grecian soil to the gods' dissatisfaction with humankind. But the gods of Hesiod were different from the gods worshiped by earlier agricultural societies. While the Sumerians, for example, believed that they worked

A nineteenth-century scroll painting shows Ainu families feasting before the staked-out body of a sacrificed bear. The Ainu—an aboriginal hunting-and-fishing people still surviving in small numbers in northern Japan—revered the bear as a sacred animal and communed with their god by eating the bear's flesh and drinking its blood. The bear's soul, they believed, would intercede with the spirit world on their behalf. Captured bear cubs were raised in cages for about two years and then killed during an elaborate ritual. At the ensuing feast, the pelt of the bear with the head attached was included in the celebration by being offered a stew made from its own flesh.

the land in order to please their fertility deities, Hesiod clearly labored only for himself, sometimes even in the face of obstacles that the gods had put in his way. Though still immortal and possessed of supernatural powers, the gods were now distinguished by human qualities: They were fickle, envious, and easily hurt. The same qualities were shared by the gods of Homer, the Greek poet of Asia Minor who composed the *Iliad* and the *Odyssey* around the eighth century BC. These gods foreshadowed a new trend in Greek thought toward a rational, more human-centered explanation of the natural world.

In one significant aspect, however, the gods of the wellborn Homer and those of the farmer Hesiod were different creatures. The former concerned themselves with prowess on the battlefield—usually the preserve of the rich, who could afford the horses necessary for heroic warfare—and were of little use to poor farmers. Hesiod's gods, on the other hand, appreciated hard work on the land, not the idleness of the wealthy elite whose well-being depended on the labor of others. This contrast reflected a developing division within Greek society between those who worked the land and those who did not.

According to Hesiod, the lords who acted as magistrates were crooked: They ground down their fellow citizens and did not fear the gods. The cause of Hesiod's own bitterness was partly personal—the judges had favored his brother Perses when the two had contended the distribution of the land that had been left to them by their father. But the complaints of the grouchy farmer probably reflected a more widespread resentment. Many of the Greek aristocrats were landowners of huge holdings, which consisted of widely scattered estates that were worked either by tenant farmers, who were obliged to surrender one-sixth of their harvest to the absentee landlords, or by landless peasants. In the *Iliad,* when Homer had his warrior-hero Achilles imagine the most wretched life he could, Achilles thought of just such an existence, as a poverty-stricken laborer.

Antagonism between the rich and the poor, mirrored in the division between city dwellers and those who lived in the countryside, attended all attempts by the Greeks over the ensuing several hundred years to make their land yield sufficient food for their growing population. By 600 BC, there had emerged a number of powerful city-states, each divided from its neighbors by high mountain ranges, and their populations soon grew too large to be fed and clothed from the meager resources of the hinterland. The problem was especially acute in Attica, a mountainous territory of more than 950 square miles jutting into the Aegean Sea northeast of the Gulf of Corinth: This region, one of the driest in Greece, whose variable rainfall during the growing season frequently caused the crops to fail, was required to support Athens, the largest city of all.

One solution to the problem was to reduce consumption by exporting the consumers. Most early cities on the heavily indented peninsula of Greece lay near the coast, within sight of the sparkling waters of the Aegean; and as food shortages became chronic, the municipal leaders looked not inland but abroad for an answer to their dilemma. Thousands of citizens were dispatched in a great wave of colonization that saw cities established as far away as Spain to the west, North Africa to the south, and Turkey and the Black Sea region to the east.

The colonists who were allotted plots of land around the new settlements were the lucky ones, for within Greece itself, the situation was worsening. One adventurer, sailing from Corinth to Syracuse in southwestern Italy, was reported to have been so hungry that he sold his promised holding for a honey cake. Because most of the productive agricultural land was concentrated in the hands of a wealthy few, and those few were increasingly planting valuable cash crops such as grapes and olives rather than grain for food, poorer farmers were forced to make use of more marginal land with worse yields. There was little opportunity for them to improve their situation. They took out loans to keep their farms going, and if their crops failed, they found themselves in debt. Often the result was that the farmers were forced to serve their creditors as virtual slaves.

Conditions in Athens became desperate, and in 594 BC, the poet and lawmaker Solon was appointed as a mediator between the rich and the poor. To ease the situation, Solon abolished all agricultural debts, allowed better-off farmers to become involved in the process of government, and banned the export of all produce from Attica except olive oil. Other states developed different answers to the problem: In Sparta, for instance, the most valuable land was held by the state and farmed by a special slave class, called Helots; in the Arcadian plain, all free citizens toiled with their own hands. But squabbles between rich and poor continued, and in some cities, the discontent was used by ambitious individuals as an excuse to overthrow the existing government and establish tyrannies, which sometimes resulted in improved conditions for the poor.

The challenge of overcoming the limitations of scarce resources was one factor that led intellectual Greeks to investigate the laws of the natural world. In the fifth century BC, the physician Hippocrates sought rational rather than divine explanations for the diseases he encountered; the historian Herodotus, whose interest in the customs and characteristics of foreign cultures was stimulated by his travels to many Mediterranean countries, came to believe that people's physical, mental, and moral qualities were largely determined by climate and environment. Such theories diminished the influence of the immortal gods and suggested that human problems could be solved only by human solutions.

By the middle of the fifth century BC, the population of Attica numbered more than 200,000 people. Their diet was by no means extravagant: Bread, which was often dipped in wine, was the staple, supplemented by goat cheese and fish and by vegetables such as beans and cabbage; most people tasted meat only on feast days. But because less than half of their land was cultivable, the citizens of Attica had become dependent on imports from grain-rich regions around the Black Sea and from Egypt, and the supply of food had become a crucial political issue. Regulations stipulated that all voyages made by merchant ships financed by Athens had to result in the importation of food, and that no person living in Athenian territory could transport grain to any port other than Piraeus, which served Athens. The penalty for the breach of these laws was death.

For the Greek city-states, the problems of food supply could be solved only by human intervention and the direct political management of natural resources. This brought new dangers, as the Athenians soon discovered: Dependence on supplies from abroad rendered the people vulnerable to attack, and when their fleet was destroyed by the Spartans in

A fifth-century-BC bronze relief portrays the squatting figure of Pan, a fertility god of the fields and forests. He is playing on a set of pipes that he fashioned from a bed of reeds into which a fleeing nymph had transformed herself. Roused from his midday slumber, Pan was lustful and unpredictable, liable to induce panic in both flocks and humans. His half-human, half-animal form personified the ambivalent feelings of the Greeks toward the natural world: Like themselves, nature could be both tranquil and unruly, and the line between what was civilized and what was bestial was never defined.

404 BC, citizens starved to death and Athens had no choice but to surrender its independence. But a pattern had been set, and the new overlords of the Mediterranean were to develop it even further.

Around 500 BC, when the Roman Republic was founded, the Italian peninsula was inhabited by a number of independent peoples, and much of the fertile land had already been tamed and put to use. Forests had been cleared around villages to provide either pasture for cattle, sheep, and goats or fields for growing grain and other crops. The Etruscan rulers of central Italy, the most powerful people in the region before the rise of Rome, had divided their estates into small plots for the use of the local population. These farmers also benefited from Etruscan land reclamation and drainage, which enabled the cultivation of areas that had previously been marshland or had been subject to flooding. The Romans had more far-reaching political ambitions than any of their neighbors, however, and in extending their territorial might, they set in motion a process that remodeled the landscape not just of Italy but of the entire Mediterranean region.

At its inception, the Roman Republic controlled an area of less than 400 square miles. By 260 BC, this domain had increased to almost 10,000 square miles, and Rome had forged alliances with other Latin states occupying an additional 40,000 square miles. At the end of the first century AD, the full-fledged Roman Empire extended from Spain in the west to Syria in the east, and from Britain in the north to the Sahara in the south. As with other empires both before and after, once the process of territorial expansion had begun, it acquired a momentum of its own. In the case of Rome, the impetus was the need to feed and clothe an ever-expanding population and to equip and pay ever-larger armies—which could be done only by gaining access to the natural resources of new provinces.

Within Italy itself, the increasing pressure on agricultural land led to changes both in the traditional system of landownership and in the uses to which the land was put. The citizens of the early republic were primarily subsistence farmers, small landholders providing for their families and trading their surplus produce for other goods. Conscripted to fight in an endless succession of foreign campaigns, many of these citizen-farmers were killed; others returned to find their farms overgrown. In due course, their land was taken over by the government, which then leased it to private individuals. Most of the best land was snatched up by enterprising aristocrats, who amalgamated select small farms into large estates. Slaves who were transported to Italy from the colonies abroad provided cheap labor, and the absentee landlords soon discovered that cash crops were a better investment than subsistence farming. As a result, the production of oil and wine for export began to take precedence over the growing of grain for local con-

A map probably prepared in the eighth century AD to illustrate a text by a Roman surveyor shows a colony founded by the Romans in 329 BC in west central Italy after the conquest of a local people. A grid of lines intersecting at right angles divides the cultivable land into squares known as *centuriae*—in theory, each square comprised 100 heritable plots. The territory was marked off by the use of a surveyor's cross, whose vertical alignment was established by plumb lines such as that on the far left, dating from the first century AD. Maps were then prepared for administrative purposes, using cartographic tools such as the folding rule and the pair of proportional compasses shown here, also from the first century AD. The gap between the lower points of the compasses is twice that between the upper points, enabling the drawing of a map at either half or double the scale of an existing one.

sumption, and much of the land was devoted to large-scale cattle and sheep ranching for meat, cheese, leather, and wool.

An agricultural treatise written in the second century BC by Cato the Elder, who himself owned several estates, gave detailed instructions on how to establish a large mixed farm for growing grapes, olives, timber, fruit, and cereal. The treatise included advice on the treatment of sick animals and on how to maintain the fertility of the soil by using the manure of domesticated animals as a fertilizer. In such practices, human ingenuity was being deliberately applied to the natural world with the aim of achieving maximum yields and profits. The new attitude was clearly expressed in a play by the Roman consul and writer Cicero, whose character Balbus the Stoic declares: "We enjoy the fruits of the plains and the mountains, the rivers and the lakes are ours, we sow wheat, we plant trees, we fertilize the soil by irrigation, we confine the rivers and straighten or divert their courses. In short, by means of our hands we essay to create a second world within the world of nature."

In the imposition of this "second world," the Romans made dramatic use of their engineering skills. Spectacular aqueducts were constructed to irrigate fields and to bring supplies of fresh water into cities. Between the fourth century BC and the early second century AD, nine aqueducts measuring a total distance of 265 miles were constructed around the city of Rome; the aqueduct at Nîmes in western Provence was 30 miles long and was lined and roofed with stone slabs. To speed the transport of fresh produce and to keep the empire's outposts in touch with the center of government in Rome, an elaborate network of roads was built across the landscape. Constructed in level terrain, these paved roads took the most direct routes to their destination—often striking straight through marshes or woodland—and were carried across rivers or deep ravines by soaring bridges and viaducts.

Engineering skills were also employed in mining gold, silver, copper, lead, and other metals. In open mines, torrents of water delivered by aqueducts—as much as nine million gallons a day at one site in northern Spain—were directed to loosen and break down deposits, and a series of sluice gates was used to process the waterborne debris. Trenches more than thirty feet deep were also cut for surface mining. To gain access to underground sources of metal, Roman engineers cut deep shafts and lined them with wood or stone; these led to a network of galleries that followed the direction of the vein to be mined. "Not only do they go into the ground a great distance," wrote the chronicler Diodorus Siculus in the first century BC, "but they also push their diggings many *stades* in depth and run galleries off at every angle, turning this way and that, in this manner bringing up from the depths the ore that gives them the profit they are seeking."

Underground lighting was provided by oil lamps placed in niches or on ledges. Water that flooded the galleries was removed either by the manual labor of thousands of slaves or, in some of the richer mines, by the water wheels or massive Archimedean screws that drew up water as they rotated—"an exceptionally ingenious machine," declared Diodorus, "by which an enormous amount of water is thrown out, to one's astonishment, by means of a trifling amount of labor."

Order and discipline were brought to bear on the surface of the land with similar rigor. In northern Italy and certain other parts of the empire, most publicly owned land was marked off by roads or paths intersecting at right angles into rectangles 2,400 Roman feet (2,300 feet) long; these plots were then divided into individual farms measuring between 25 and 100 acres. Villas—essentially farmsteads of above-

average size and sophistication, often containing facilities for oil and wine production, weaving, dyeing, and metalworking—replaced more basic agricultural settlements around major towns in the provinces. And after the North African city of Carthage was captured about 200 BC, the wide belt of fertile soil between the Mediterranean coast and the Sahara was rapidly exploited to produce grain for export to Rome. Much of the land was divided into a grid system of 125-acre squares, which were settled and farmed by military veterans. Woodland was cleared, and pastures and orchards were plowed under. "Cultivated fields have become the forests," observed the Carthaginian theologian Tertullian. "The sands are being planted, the rocks hewn, the swamp drained."

The timber from the forests that yielded to the plow was itself a highly prized commodity. Wood was virtually the only fuel available for heating, cooking, and the smelting of metal ore; it was used extensively for constructing houses and making furniture; and, above all, it sustained military conquest. The Roman natural historian Pliny the Elder recorded that during the first war against Carthage, a fleet of 120 wooden ships was built in sixty days, and for another naval campaign, 220 ships were built in just forty-five days. They were destroyed almost as quickly: According to another Roman historian, 700 *quinqueremes*—massive galleys with five banks of oars—were lost in that first war against Carthage. Armies were even more voracious in their appetite for timber: In addition to being used for fuel, wood was employed for building bridges, roads, defense works, and siege towers. One tower that was erected to storm the walls of Rhodes was described in detail by a chronicler: Nine stories high, its top platform was 150 feet above the ground, and it required a force of more than 1,000 men to move it into position.

Meanwhile, as the empire expanded, so too did the city of Rome: By the first century AD, the population of the capital had grown to more than one million. In 123 BC, the civic authorities had taken on the responsibility of distributing a free ration of wheat every month to all eligible citizens; slaves and those who did not qualify for the handouts were supplied by private merchants. The poorer inhabitants of Rome had, therefore, come to expect guaranteed supplies of food, and satisfying this demand became a political problem of increasing urgency. Around AD 50, the emperor Claudius was attacked in the Forum by a hungry mob who pelted him with pieces of stale bread. Recognizing that this was a situation as volatile as any on the empire's remote frontiers, Rome's rulers determined to buy off discontent with cheap bread and cheap thrills. Both elements in this policy of bread and circuses intensified what had now become a coordinated assault on the natural resources of the provinces. To keep the bellies of Rome's citizens full, wheat had to be supplied quickly and in bulk. And to take their minds off rebellion, animals—the wilder and the more exotic the better—had to be captured for bloody gladiatorial spectacles.

Most of Rome's wheat was supplied from Sicily, southwestern Spain, and North Africa. Their distance from Rome was not a great impediment: Despite the hazards of storms and piracy, it was cheaper and quicker to transport grain by sea than by overland route. But the drain on the agricultural resources of the provinces proved so great that many of their own inhabitants were left without sufficient food. Rome was consuming more than 6,500 tons of wheat per week at the beginning of the first century AD. To keep pace with this demand, at least 800 shiploads of wheat had to reach the port of Ostia on the Tiber River during the sailing season, which generally lasted from May to September. During the following century, the Greek physician

This third-century mosaic from a Roman villa in present-day Algeria shows frantic leopards and lions being driven toward a net that is securely attached to stakes. While advancing on the animals, hunters push flaming torches from beneath their shields; on the left, one reckless pursuer has been caught and felled by a leopard. The purpose of such hunts was to capture the animals live for transportation to Rome, in whose thronged arenas they would be pitted against gladiators in bloody, spectacular contests. The local extinction of the elephant, the rhinoceros, and the zebra in North Africa and the hippopotamus from the lower Nile region in Egypt was largely a result of the Roman craze for imported exotic beasts.

Standing at the helm of his boat, a captain supervises the loading of a cargo of grain in this third-century mosaic from Ostia at the mouth of the Tiber River. In the first century AD, about 330,000 tons of grain were carried each year to Ostia by seagoing freighters, chiefly from North Africa and Egypt; from the port's warehouses, a shuttle service of lighters sailed twenty miles upstream to Rome. The coin adorned with six stalks of wheat *(inset)*, dating from 20 BC, commemorates the reorganization by Emperor Augustus of free supplies of grain for Roman citizens. His successor Tiberius warned the Senate that if this dole were discontinued, the "utter ruin of the state will follow."

Galen complained that the "city dwellers, as it was their custom to collect and store enough grain for all the next year immediately after the harvest, carried off all the wheat, barley, beans, and lentils and left what remained to the countrypeople." The latter finished off their beans during the winter "and so had to fall back on unhealthy foods during the spring; they ate twigs and shoots of trees and bushes and bulbs and roots of indigestible plants; they filled themselves with wild herbs and cooked fresh grass."

Animals were also imported from abroad, and in such numbers that the larger species became extinct in regions accessible to professional hunters. The wealthy maintained a ruling-class tradition of keeping certain animals as pets: The emperor Hadrian erected tombstones over the graves of his favorite dogs, and many aristocrats kept monkeys, birds, and even fish as domestic playthings. But as far as ordinary Romans were concerned, animals were for providing meat and milk, hauling wagons, dragging plows—and for entertainment. Before massed crowds in the Circus Maximus and other stadiums, bears, elephants, lions, leopards, crocodiles, and rhinoceroses were slaughtered by professional fighters or goaded into destroying one another, having first been drugged with intoxicants to heighten their frenzy. Beasts that survived were shot with arrows by spectators who had paid extra for their ringside seats. The emperor Nero had 400 bears and 300 lions killed with javelins by his personal bodyguard. To celebrate his military victories, Emperor Trajan had 11,000 animals publicly butchered. Sensible citizens were appalled: "Can't you people see even this much, that bad examples recoil on those who set them?" asked the philosopher Seneca. But the policy served its immediate purpose, and the populace of Rome, sated, went peacefully to bed.

Warfare and agriculture, the twin engines of the empire, fed ceaselessly upon each other. Since the large estates in Italy, Sicily, and eventually, in North Africa were also worked by foreign slaves—the great markets of Capua and Delos were capable of handling 20,000 slaves per day—the dispossessed small landholders could not even sell their labor. They therefore drifted to the urban centers, increasing the demands that each city placed on its hinterland to provide for its citizens. Moreover, the ready availability of slaves discouraged the development of new agricultural techniques—so that output could be increased only by further military conquest, leading to an extension of the area under cultivation.

But this apparently endless cycle in fact contained its own constraints. First, there were the limits of human skill: The Romans proved unable to adapt their high-yield agricultural methods to the wetter climate and heavier soils of central and western Europe. Second, there were the limits of the land itself. Cleared, mined, and constantly plowed, the soil began to lose its fertility. Around AD 250, the Christian bishop of Carthage sent a warning to the Roman proconsul in Africa:

You must know that the world has grown old and does not remain in its former vigor. The rainfall and the sun's warmth are both diminishing; the metals are nearly exhausted; the husbandman is failing in his field . . . springs that once gushed forth liberally, now barely give a trickle of water.

For those with keen eyes, the evidence of this decline had been there to see at least 300 years earlier. The deforestation of the Italian Apennines during the wars against Carthage had allowed soil to be washed down the hillsides, silting up rivers and turning coastal areas into marshland—including the Pontine Marshes at the mouth of the Tiber River. Around 60 BC, the poet Lucretius observed that farmers were having to work harder than their ancestors to harvest the same amount of produce. But it was in North Africa and Asia Minor, the regions of the Mediterranean with the lowest

This third-century-BC Greek terra-cotta model of a camel bearing amphorae of wine was found in Syria, which exported grain to the populous city-states of Greece and received in return Greek wine, oil, and pottery. The fastest and cheapest method of transporting produce was by ship, but along caravan routes traversing the hot, dry regions of Asia and the Middle East, the camel came into its own. The animal's ability to store fat in its hump and to conserve body water, as well as its high tolerance of dehydration, enable it to travel for many days on a sparse diet of thorny bushes or dried grass. These attributes made the camel an ideal pack animal for the desert plains of North Africa, where it was introduced about the third century AD.

rainfall, that the effects of overexploitation of the land were most conspicuous. Here, lush forests had once absorbed the precious winter rains and retained moisture in the soil around their roots. Now, with the trees gone, the soil exhausted, and supplies of manure too little or too distant to be brought in to replenish the land's fertility, the wind whipped away the dry topsoil, and the region began to turn to desert.

The Romans did take positive steps to conserve their resources. Marshland was drained in Italy, and terraces were carved in the hillsides of North Africa to hold rainfall or melting snow, which seeped slowly through the soil to fill the water table and replenish wells. In Rome in the first century AD, there was a lively market for recycled glass: Citizens collected fragments from the streets and sold them to local workshops, where less fuel was required to recycle the old product than was needed to make new glass from raw materials. At Ephesus, the major city of Asia Minor, where alluvium washed down from the hills by the Cayster River constantly threatened to silt up the port, the Romans repeatedly dredged the harbor and even attempted to divert the course of the river. But all such measures provided only a temporary alleviation of the problems.

Moreover, there were signs that the organizing genius and energy of Rome had begun to wane. "No addition whatever is being made to knowledge by means of original research," complained Pliny the Elder, "and, in fact, even the discoveries of our predecessors are not being thoroughly studied." Many ordinary citizens, especially among the military and lower classes, were dismayed by the manner in which the official Roman religion had become deeply intertwined with political protocol—some emperors even decreed that they themselves were gods—and switched their loyalties to charismatic mystery religions that flooded into Rome from the East. Among these were the cult of Isis, a version of sacrificial-bull rites that had flourished in Egypt centuries before, and a cult derived from the ancient Indo-European deity

A page from an eleventh-century Arabic edition of a Greek catalog of some 600 plants and their medicinal uses shows the agrimony plant, which was used to cure liver ailments. The inventory was compiled in the first century AD by Pedanius Dioscorides, one of a number of Greek scholars who rejected magical and divine explanations of diseases in favor of empirical observation of symptoms and cures. Subsequently translated into at least seven other languages, Dioscorides's work served as the primary text of pharmacology in Europe until the end of the fifteenth century.

Mithra, which also involved the sacrifice of bulls. A third contender for the allegiance of Romans who turned away from their traditional gods was Christianity, which despite fierce persecution made great strides among the populace. In the fourth century, Christianity was adopted as the official religion of the empire—a dangerous moment, for the new faith attacked the class system and rigid social hierarchy, built to a large extent on slave labor, on which the fortunes of Rome were based.

There were dangers from without, too. For centuries, the Romans had been menaced by nomadic raiders who lived on the open plains across the Danube River and preyed on the northern frontier. *Barbari,* the Greeks had called them, lampooning their incomprehensible, barking language. In the fourth century, these peoples began to advance south and west, swarming over the imperial defenses and into France, Spain, Italy, and North Africa.

Rome in its senescence, ruled by a succession of weak emperors and succumbing to a mood of fatalistic pessimism, could offer scant resistance to the invaders. In AD 476, Rome's control over the western empire, for some time hardly more than nominal, was officially terminated.

The inhabitants of the entire Mediterranean region and western Europe gradually came to realize that an era had ended. Merchant fleets, packed tight with grain, no longer converged on the wharves of Ostia. Mine shafts caved in. Water ran dry in aqueducts, and weeds sprouted in the precincts of abandoned temples and on roads once trodden by the marching feet of legions. To many witnesses it must have appeared as if the land itself, along with the power of Rome to which it had been harnessed, had grown old and impotent.

Even more gradually, however, the land began to reassert its own rhythms of growth, as if to prove that all along it had retained its essential independence—and though borrowed by Rome and put to Roman uses, it was itself no more Roman than it had been Minoan, Mycenaean, Greek, or Phoenician. Some regions, however, would never return to their former fruitfulness: The more eroded parts of North Africa became home not to settled farmers but to nomadic pastoralists, whose herds of goats and other foraging animals prevented new tree cover from growing. But most parts of the Mediterranean area, no longer subject to the overweening demands of Rome and its provincial capitals, now had to supply more immediate wants, and proved equal to that challenge. Rivers still flowed in a seasonal cycle of spate and decline; olives, vines, wheat, barley, and a variety of vegetables nourished the people of villages and small towns.

Lapsed from the severe and uniform regimen of the Romans, the resilient Mediterranean landscape became a mosaic of varied and productive local agricultures—and would remain so until the technology and mass demands of the twentieth century imposed their own exacting monocultures. Even these did not trespass—for by then there was profit in tourism, as well as a judicious sense of limits—on the hollow stone ruins of Greece and Rome.

THE CHINESE WAY

3 During the reign of the legendary Chinese emperor Yao, ten suns shone in the sky at the same time, causing rivers to dry up and the earth to become barren. By dint of blind perseverance and determination, the hapless people of this drought-stricken land managed to eke out a meager livelihood—but just as they were beginning to prosper, a great flood swept over the world and reduced them again to misery.

So begins an account dating from several hundred years BC of the beginnings of Chinese civilization. But the story goes on to tell how, with divine help, the destructive forces of nature were eventually tamed, and order and control were brought to the land. A god named Gun took pity on the suffering people and, using a piece of soil with magical properties, built a dam to hold back the floodwaters. Gun was killed by the angry god of fire, and again the waters covered the land. But Gun's spirit refused to die, and after three years, a son, Yu, emerged from his corpse. Yu was able to obtain the services of a winged dragon and also a new supply of the magical soil, which he used to stop up the 233,599 springs from which the water was emerging.

Yu next turned his attention to the floods that still covered the landscape. For thirteen years, he traveled by boat on water, by sleigh over marsh, and by cart on land; he did not stop to enter his own house even though he passed by the gate. In this manner, he surveyed the land and marked off the boundaries of nine provinces; he also cut deep channels that became the great rivers of China, made passes through the mountains, and built dikes in the low-lying plains. Yu also fathered a son, Qi, who became the founder of the first hereditary dynasty in China.

Floods and drought have repeatedly punctuated the course of Chinese history, and Yu's efforts to bring order and stability to this troubled land speak for the experience of countless generations of ordinary Chinese people. Their successes have always been precarious: Famine has so often canceled out the labor of hundreds of thousands of families that for most Chinese, for most of the time, bare subsistence has been the only realistic goal. But their achievement, however hard won, is undoubted. The Chinese now have a life expectancy—sixty-eight years at birth for males and seventy-one years for females in 1988—that is higher than that of the U.S.S.R., India, and any nation in Africa, and comparable to that of the richer nations of the Western world, which are themselves less subject to natural calamities.

In coping with the natural disasters that beset their land and also with the economic problems of feeding an enormous population, the Chinese and other peoples of eastern and southern Asia have resorted to methods of agriculture very different from those characteristic of the West. From at least the beginning of the nineteenth century, America and Europe have fed their own citizens by the evermore intense application of industrial technology. In China and India, other strategies—determined by the land itself and by cultural attitudes toward it—have prevailed. The effects have been most

A detail from a sixteenth-century Japanese screen shows a team of women planting rice in a flooded field, where the seedlings will remain until the field is drained in preparation for harvesting. Wet cultivation of rice was established in China possibly by the end of the fifth millennium BC; around 500 BC, it spread to Japan and also to India and many parts of Southeast Asia. Rice growing, though one of the most highly labor-intensive forms of agriculture, is also one of the most productive: Rice today is the chief food for three out of five of the world's inhabitants. In many Eastern countries, handfuls of rice are thrown over newly married couples to wish them fertility, a custom from which the Western practice of scattering confetti derives.

apparent in the biological sphere: the introduction of new varieties of crops, many of them developed for growing on inferior land, and of labor-intensive systems of agriculture that have made efficient and economical use of natural resources. China and the Orient have prospered not so much by mastering nature as by the patient expenditure of infinite amounts of human energy and minute attention to detail.

Present-day China is the third-largest nation in the world and home to more than one billion people, one-fourth of the earth's population. But such impressive statistics conceal an immense variety of terrains, of peoples, and of cultural traditions. If China

The hand of the farmer has transformed this landscape near Yangshao in southern China into a patchwork of fields interlaced with irrigation canals. Because of the pressure of supporting an enormous population, every cultivable piece of land up to the edge of the sheer-rising mountains is used to produce crops: rice in the flooded paddies, and wheat, beans, vegetables, and tobacco in the dry fields.

exists at all as a unified entity, in the minds of both outsiders and the Chinese themselves, it is only because of a long historical process that involved the development and implementation of complex and resilient forms of political control; an eclectic but unifying culture; and methods of working the land that rely above all on cooperation and shared endeavor.

Astronauts orbiting the earth have reported that the only construction visible from space is the Great Wall of China, the barrier erected along the northern frontier to protect the settled Chinese from their nomadic, aggressive neighbors. As they tumbled toward the earth and the surface of the globe came into sharper focus, the astronauts might have noticed that China comprises two distinct geographical regions. Inner China, forming a rough square bordered to the south and east by the sea, is a land of alluvial plains and river valleys, a patchwork of tiny fragments of every shade of green and brown. The historical core of the Chinese empire until the seventeenth century, this terrain is home to about 95 percent of the population. On its northern and western sides, it is wrapped around by outer China, a region of greatly contrasting topography. Outer China includes the world's highest mountain—Mount Everest, towering 29,028 feet above sea level—and the second lowest point on the earth's surface, the Turfan depression, 426 feet below sea level. It encompasses the burning deserts of the Gobi and Takla Makan; the rolling, grass-covered steppes of Mongolia; and vast tracts of swamp and virgin forest. Not surprisingly, large areas are uninhabited, and the overall population density does not exceed 2.6 people per square mile. Recent centuries have seen the migration of substantial numbers of people from inner China into parts of outer China, notably Manchuria, but these peripheral regions are still primarily the homelands of the Mongols, Tibetans, and other peoples who make up fifty-four of the fifty-five ethnic nationalities recognized by the People's Republic of China.

Simple human habitations nestle by a lakeshore in this scene from a scroll painted in ink by the seventeenth-century Chinese artist Wang Hui. Rejecting realism, traditional Chinese painting favored harmonious, idealized compositions in which the pulse of nature was expressed in the forms of lofty peaks, surging waterfalls, and branching trees. Such works embodied reverential attitudes toward the natural world that coexisted with its necessary and increasing exploitation.

Inner China in turn can be divided into two distinct areas. From the mountainous region to the west, the Qinling and Daba ranges of the Kunlun Mountains extend like a finger and thumb to grab the trailing line of the Han, a tributary of the Yangtze River. This river and the mountains it threads into define the boundary between very different climatic and geographical zones.

South China—as also the whole of Southeast Asia— ranges from the temperate to the subtropical in climate. In the Sichuan basin and toward the South China Sea, temperatures and rainfall are high, and the growing season lasts nearly year-round. In ancient times, the land was covered by dense forests of pine and bamboo, home to creatures such as elephants and giant lizards; the human inhabitants occupied the coastal lowlands and valleys and supplemented their simple agriculture with hunting and fishing. Farther north, in the Yangtze Valley, rainfall is lower, and there is a tendency toward dry winters. Summers are hot and long, and the growing season lasts nine to ten months. The first-century historian

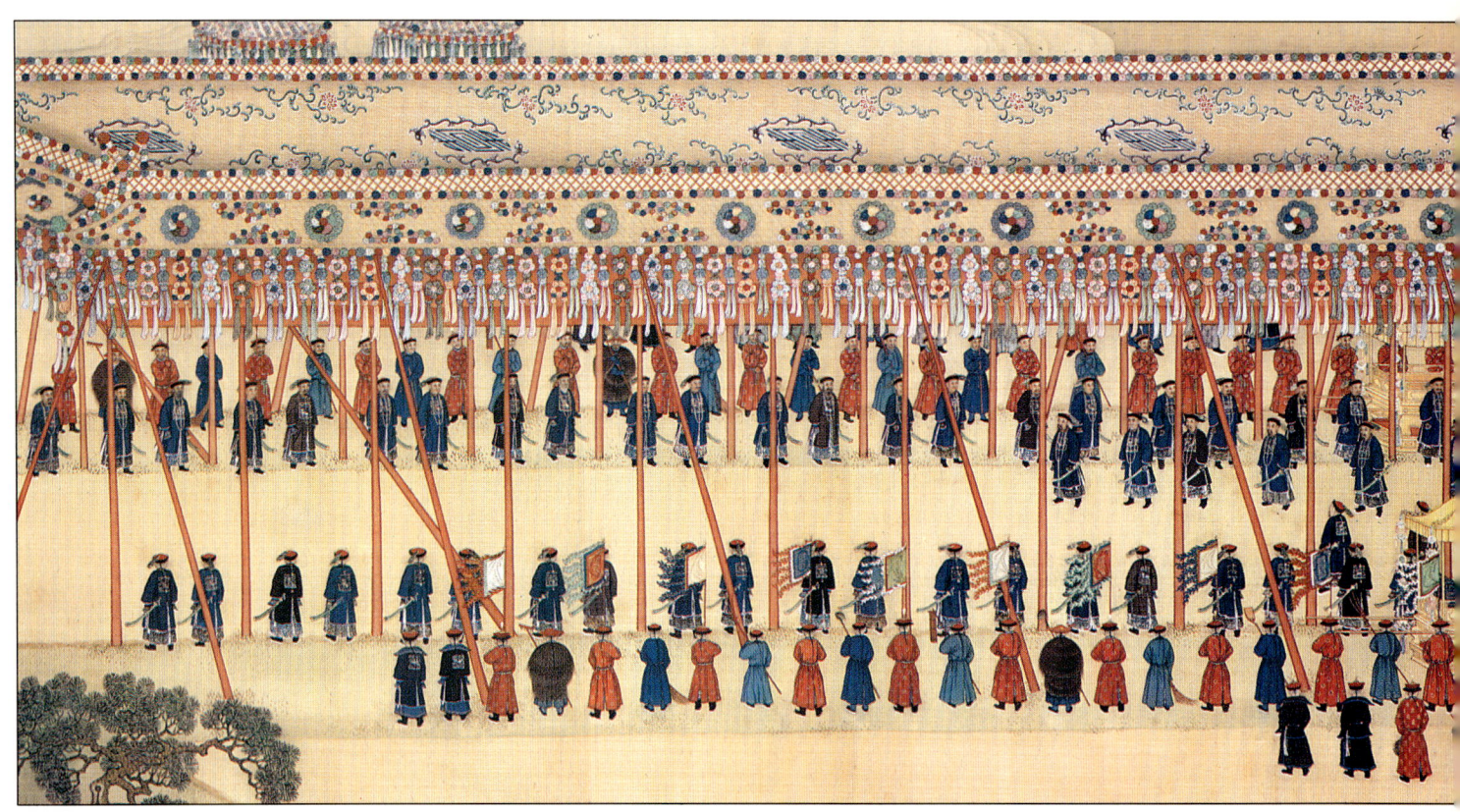

An eighteenth-century silk scroll shows ranks of courtiers under a lavish canopy attending the emperor—descending from the dais on the far right—who is about to plow the first furrow of the year. Parallel with the emperor and awaiting his lead, high ministers stand behind their own plows. In addition to performing this ceremony, intended to ensure a successful growing season, Chinese emperors since the earliest dynasty were expected to offer a sacrifice each spring and autumn at the altar of the spirit of the soil. These customs suited imperial interests: Good harvests lessened the likelihood of rebellion, and most of the emperors' revenue derived from taxes on land and produce.

Ban Gu wrote an observant if subjective description of the lifestyle enjoyed by the denizens of this region during the fourth century BC:

> The people live on fish and rice. Hunting, fishing, and wood gathering are their principal activities. Because there is always enough to eat, they are a lazy and improvident folk, laying up no stores for the future, so confident are they that the supply of food and drink will always be replenished. They have no fear of cold and hunger; on the other hand, there are no rich households among them. They believe in the power of shamans and spirits and are much addicted to lewd religious rites.

To the north of the Yangtze lie wide plains through which the Yellow River, the second of China's two great arteries, passes. The plains are covered by a fine, yellow loess that was picked up by winds from the deserts of central Asia during the end of the glacial period and deposited over the northwest of China, sometimes to a depth of several hundred yards. Loessial soil is easily worked and fertile, and about 500,000 years ago, it would have been pierced by outcrops of forest-covered rock, a source of both timber and water. Here, where the Yellow River emerges from the mountains but is still protected by hills, the first Chinese people lived.

As early as 5000 BC, communities of farmers were living in villages in what are now Shaanxi and Henan provinces in northern China and in the settlement of Longshan, farther to the east in modern-day Shandong province. Their houses had walls of wattle and daub and floors of beaten earth. They had domesticated dogs and pigs and had developed sophisticated techniques of making and decorating pottery. Using

implements of stone and wood, they cultivated millet, wheat, beans, rice, hemp, cabbage, and melons. Gradually, agricultural prosperity led to the development of cities, the first of which were probably established by clans headed by warrior-leaders whose dominance was ensured by their possession of bronze weapons and horse-drawn chariots. The Shang kings of the seventeenth to the twelfth centuries BC indulged in great hunting expeditions as training for their endless forays against other cities and against less-developed peoples occupying the surrounding countryside; the remains of elephants, rhinoceroses, bears, tigers, leopards, deer, monkeys, foxes, and badgers have been found on their hunting grounds. The staple crops of wheat and millet supported a growing population, and the kings were able to call on the services of scribes and other officials who were not involved in the production of food.

As the use of bronze technology expanded, peripheral regions were settled and brought under cultivation, and the increasing grain production supported larger armies. One by one, smaller and weaker states were subsumed until, in 221 BC, the state of Qin commanded a unified Chinese empire for the first time. The Qin state sought to establish its authority with systematic vigor: Texts deemed subversive were proscribed; weights and measures were standardized; travel was allowed only with a permit; and armies of corvée labor were drafted to undertake massive irrigation and road-building works. But the death of the first emperor was soon followed by rebellion, and in 206 BC, a new dynasty, the Han, was established. The next 400 years of the Han dynasty saw the expansion of Chinese culture to a large part of inner China.

Great numbers of people had already migrated south, fleeing wars or natural disasters that threatened the stability of life in the Yellow River homelands. Around the fourth century BC, more than one-third of the registered population of North

China, approximately two million people, had moved south of the Yangtze; and in the wake of mass migration to well-endowed regions such as the Sichuan basin, people were further displaced to the south and southwest. The Han rulers encouraged such population movements: Seeking to maintain control of the trade routes to the West and to central Asia, they established self-sufficient colonies of farmer-soldiers in the fertile river valleys and settled as many as 700,000 colonists in the Gansu corridor in the northwest at the end of the second century BC. Those peoples who did not submit to Chinese rule were pushed out of the cultivable lands and forced to adopt a pastoral, nomadic way of life on the steppes beyond.

Trade filled the Han coffers to overflowing: The Roman Empire's insatiable demand for silk and Chinese spices ensured a continuous stream of merchants along the caravan route through central Asia known as the Silk Road, the balance sheet being very much in favor of the East. Agriculture, too, brought a new measure of stability. Zhao Guo, an official at the Han court during the first century BC, recommended plowing wide and deep furrows; when the seedlings sprouted, the soil from the ridges was piled around them, until the field was level. Such methods made harvests more secure and were put into practice throughout the empire.

At the same time, a diverse range of intellectual and religious influences was shaping distinctive cultural attitudes toward the natural world. Among these influences were the ancient concepts of yin and yang—complementary but opposing forces that represent darkness and light, female and male, weakness and strength. Yin and yang were believed to rise and fall cyclically, a theory that was used to explain the pattern of birth, death, and rebirth in the natural world as well as the progression of the seasons. All human and animal activity was believed to have its optimum point in the cycle, and to carry out an activity outside the appropriate time spelled disaster. The influence of yin and yang could be channeled by judicious human intervention, however. In his account of his travels in China during the ninth century, the Japanese monk Ennin reported the following traditional method of controlling the climate: "When, seeking good weather, you block the north road, yin is obstructed and yang then pervades and the skies should clear. When, asking for rain, you block the south, yang is obstructed and yin then pervades and rain should fall."

The teachings of the philosopher Confucius, who lived in the sixth century BC, provided justification for the exercise of authority both in the state and in the family by concentrating on form and ritual. Ideal standards of conduct for everyone from rulers of state to children were defined, and exemplary individuals were held up as models. From a Confucian point of view, natural disasters and political downfall were both seen as the result of a failure to observe the correct rites.

A very different way of understanding the world counterbalanced the somewhat conformist doctrines of Confucius and his followers. Laozi, who was believed to have lived during the same era as Confucius, and his successor Zhuangzi were interested in attaining freedom from man-made restrictions and rules, hoping thereby to achieve enlightenment—that is, an understanding of the Dao, or Way, the principle underlying the universe and all creation. Those in search of the Dao were often inspired to become hermits, finding a solitary life in a mountain retreat more conducive to contemplation on such matters than the hurly-burly of society. Such a life was portrayed by generations of Chinese painters in their pictures of mist-shrouded mountains on which the tiny figure of a white-haired scholar gazes from the shelter of his rustic pavilion. The fifth-century writer Kung Zhigui described the ideal toward which

Daoist recluses should aspire: "A man who, incorruptible, holds himself aloof from the vulgar; untrammeled, avoids earthly concerns; and vies in purity with the white snow, ascends straight to the blue clouds."

Later Daoists claimed to have received revelations of sacred texts, and Daoist cults devoted to religious healing achieved a popular following. Seeking to escape the penalities of mortality, some Daoists put their interest in natural phenomena to use in the quest for the elixir of life—although for those whose chemical experiments made use of poisonous materials such as cinnabar, this search sometimes brought about the exact opposite of the desired result. Daoist priests, based in temples set in remote parts of the countryside, collected and categorized herbs and medicinal plants, compiling important pharmaceutical natural histories. Often, they noted surprising—and unproved—associations between natural phenomena: "When in the mountains there is the *xia* plant, a kind of shallot," according to a Daoist treatise written about AD 800, "then below gold will be found. When in the mountains there is the ginger plant, then below copper and tin will be found. If the mountain has precious jade, the branches of the trees all around will be drooping."

Daoist priests became practiced in the art of geomancy—known in Chinese as *fengshui,* or "wind and water"—which ensured that tombs and buildings were not sited in locations that disrupted the powerful forces believed to run through the landscape. Despite offical disapproval, geomancers still operate in rural China.

A third body of thought to make a lasting contribution to Chinese attitudes toward their world was of foreign origin. Buddhism was introduced to China in the first century AD by Indian missionaries following the traders on the Silk Road. It was first adopted by the upper echelons of Chinese society and underwent substantial transformation as the belief became more popular. In China, for example, the Buddhist

PUTTING WATER TO WORK

Both blessing and curse, China's waterways nourish agriculture and—by flooding adjacent plains at irregular intervals—threaten mass devastation. Since at least the third century BC, armies of laborers have toiled to construct irrigation canals and flood-proof dikes; their largest project was the building in the seventh century of the Grand Canal, shown overleaf, whose main function was to transport grain over a distance of about 620 miles from the south to the north. A detail on the right from a seventeenth-century scroll shows workers bringing a tied bundle of reeds to repair a dike while others hurry by carrying mattocks and baskets of earth. Some overseers offered prizes of meat, wine, or clothes to the fastest workers.

concept of the transmigration of souls was developed into ideas of paradise and the underworld that were quite alien to Buddha's original teachings. But the promise of peace and plenty after death had a widespread appeal to a population whose earthly lives were both exhausting and perilous. An ancient adage neatly summed up the pragmatic attitudes of the Chinese toward their eclectic cultural traditions: "Confucian in office, Daoist in retirement, Buddhist as death draws near."

The first Chinese census was conducted in the year AD 2 and numbered the population of the Han empire at around 60 million. The borders were still insecure, and the people within the empire followed many different ways of life, but the very fact that the census could be taken indicated that a strong central authority had been established over a vast area. A major contribution to this unity was the degree of communal effort required to bring the land under human control and make it bear sufficient produce for a growing population.

The first imperative was to harness the potentially destructive torrents of the major rivers. Vast armies of laborers were mobilized to erect dikes to confine the waters of flooding rivers and to dig ditches to divert this resource for soil irrigation. All the major rivers of China, the Yellow River in particular, pick up large amounts of soil as they cut through the hills and mountains in their upper reaches. This load is carried downstream until, as the flow of the river slows upon reaching the plains, it is deposited on the riverbed, thus raising its level and causing the water to burst its banks. The Yellow River, known as China's Sorrow, carries some 1.6 billion tons of silt annually and has changed its course dramatically during the past 2,000 years—

from following a path almost due north, to debouching south of the Shandong Peninsula, and then moving back again to emerge where it does today in the Bo Hai gulf. In some parts of the alluvial plain, centuries of dike building have resulted in the river's being raised many feet above the surrounding landscape.

A complex system of irrigation canals in the low-lying delta both evenly distributed water over the land and, by means of reservoirs, enabled water to be stored for times when rainfall was scanty. The building of irrigation channels and dikes had a further effect on the soil of the delta region: The silt dredged from the canals during periodic clearing was deposited on the fields, creating a layer of topsoil that sometimes reached depths of more than three feet.

The most ambitious project designed to put water to productive use was the construction of the Grand Canal at the beginning of the seventh century AD. The canal's primary purpose was to enable the rice grown in the Yangtze Delta region to be transported quickly and cheaply to the capital in the north and to the soldiers stationed on the northern frontiers; because of the absence of a good road network, especially in the southern part of the country, it was also used for many other kinds of traffic. The canal began in the fertile fields of the Yangtze Delta and then crossed the low-lying Huai River valley, incorporating conveniently placed tributaries in its path; a northeastern arm reaching as far as present-day Beijing served the garrisons there. It was built by a conscript force of 5.5 million laborers supervised by 50,000 guards, and in the ensuing centuries, workers were drafted by the hundreds of thousands for periodic repairs.

Raising water from canals to irrigate the fields required the use of pumps, powered by the wind, animals, or more usually, people. Simple devices such

  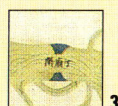
1　　　　2　　　　3

A section from a forty-six-foot-long scroll map of the Grand Canal *(left)* drawn in the eighteenth century depicts the junction of the canal with Lake Luoma and the Yellow River and the beginning of its southward course to the Yangtze River. That the canal is shown here to run parallel to the Yellow River *(colored yellow)* is the result of a distortion of directional values that allows the map to include an abundance of information of benefit to contemporary administrators. The boxed details above show symbols representing (1) dikes to prevent flooding; (2) the placement of gabions, the stone-filled bamboo cages used to repair dikes; and (3) locks in the canal. Also recorded here are mountains, springs, towns, and bridges. The passages of writing are lists of villages and families responsible for the maintenance of specific dikes.

as the well sweep, a counterbalanced bucket, and treadmills, which consist of a chain of buckets or paddles that carry or push before them a supply of water, are still in use in many parts of China. A more sophisticated device introduced in the tenth century is the noria, a form of giant waterwheel whose highest scoops spill water into a pipe that is carried on props over the adjacent fields. Where a fast-flowing river is the source of power, the noria can be self-propelling. A twelfth-century poet considered these devices worthy of praise in verse:

> Ten tubes upon each wheel both drain and irrigate,
> Rising and falling in a circle without cease.
> Rich lands have been opened up in the surrounding hills,
> Along ten thousand acres wind rice-sprouts like green clouds,
> Clouds that before your eyes turn from green to gold,
> And fill all stomachs with the year's rich harvest.

The crop celebrated by the poet was grown mainly in the south, where rainfall conditions were ideal, and as this region became more populated, rice gradually supplanted wheat and millet, the more ancient staples of northern China. Indeed, it was to become the most plentiful crop of the entire world: Approximately one in twenty-five of the earth's arable acres is now used for cultivating rice, which provides more than 650 calories per day for every man, woman, and child.

The highly efficient system of air passages leading from the leaves to the root of the rice plant enables it to grow in waterlogged conditions that would drown most other crops. Some varieties of rice can survive under floods up to sixteen feet deep, growing almost ten inches a day to keep their leaves above the rising waters. And far from being a handicap that the plant has had to overcome, water in fact promotes its growth: Controlled flooding both prevents nutrients from escaping into the atmosphere or the subsoil and makes for a highly efficient supply of nitrogen and phosphorus, the two most important plant minerals. Without the aid of artificial fertilizers, well-tended rice fields can produce two or three harvests each year for centuries from the same area of land.

Rice growing was and remains a highly labor-intensive activity. There are no mechanical devices that can perform the tasks of sowing the seeds in prepared beds, transplanting them to paddies after about one month's growth, and then weeding and harvesting them with the same efficiency as human hands. When traditional methods of cultivation are employed, every few acres of land require between 1,000 and 2,000 hours of labor to produce just one harvest. In addition, a high degree of human organization is needed to control the use of water in the rice paddies and on the terraces that have been cut into hillsides: Because the land must be dry when the rice is harvested and remain so for a short time afterward, growing cycles must be staggered in order to ensure continuous productivity. The amount of labor and coordination required, however, have themselves contributed to the stability and enduring success of this form of agriculture.

Where conditions determined their need, further sophistications were built into the system. If the growing season was not long enough to allow successive harvests, intercropping was practiced, in which a second planting of seedlings was set out in alternate rows a few weeks after the first. And where land suitable for cultivation was in short supply, different types of plants that could be harvested without mutual

On a bamboo lattice, silkworms that have been fed on mulberry leaves spin the fine silk fiber secreted by their glands into shell-like cocoons. The candor of this painting, produced for export in the nineteenth century in a style designed to appeal to European tastes, contrasts with the jealousy with which China guarded its monopoly on silk production for more than 2,000 years. By the eleventh century, sericulture—the art of raising silkworms—had spread to Japan and, via India and the Arabian Peninsula, to Europe.

interference were interwoven in rows. A Han dynasty text recommended growing scallions under melon vines, and other works suggested combining onions and coriander, and turnips and hemp.

To ensure that the yields generated by such intensive farming stayed high year after year, the Chinese peasantry developed elaborate methods of maintaining the fertility of the soil. Organic material dug into the soil to improve its nitrogen content included leguminous plants such as green beans, mud taken from rivers and canals during dredging operations, animal dung, and night soil—that is, human excrement. Farmers augmented their own family's production of this vital resource by building latrines at the side of the road for the use of passers-by, to mutual benefit. In cities, night soil was collected each morning and transported out to the countryside either to be applied directly to the fields or to be dried for later use. The sight of euphemistically named "honey wagons" being pulled along city streets by horse or by people is still a common one. The cumulative effect of the nurturing of the soil over generations is evident in the high humus and nitrogen levels of the zones around cities.

In addition to the staple cereal grains of wheat and rice, the Chinese peasantry grew a range of other crops. A guide for the improvement of rural life written in the sixth century AD gave advice on the cultivation of vegetables, fruit, fiber crops, and trees for timber; it also included recipes for food, medicines, and wines as well as instructions on how to make such daily necessities as soap, glue, and dyes. The treatise stressed the importance of being guided by the weather and the type of soil available: "Following the appropriateness of the season, consider well the nature and conditions of the soil; then and only then, least labor will bring success. Rely on one's own idea and not on the orders of nature, then every effort will be futile."

The tensile grace of bamboo stems—celebrated in the poem inscribed on this thirteenth-century ink drawing—is just one of the plant's virtues. Ubiquitous in China and widely grown in Southeast Asia and on the islands of the Indian and Pacific oceans, bamboo grows faster than any other plant, up to three feet a day. Cooked young shoots are eaten as vegetables; the pulp and fiber are processed to make paper; and the chemical compounds contained in some species are used in medicine. Bamboo is also used to make a vast range of domestic artifacts as well as to construct houses and build ships. Shown above, for example, is a section of a 755-foot-long bridge over the Min River in Sichuan province that is made entirely of bamboo.

Subsidiary crops have included corn, sweet potatoes, kaoliang—a type of sorghum—and peanuts. Tea bushes were grown on hillsides throughout South China, where variations in soil and climate in different districts produced distinctive flavors. Soybeans proved a rich source of protein and were also important in crop rotation to improve the soil. Hemp and ramie, a type of nettle, were the plants from which the clothes worn by most Chinese were made until the fourteenth century, when cotton was introduced from India and Southeast Asia.

For centuries, China's most profitable export was silk, which also—as a woven fabric or as wadding for padded garments—clothed the Chinese upper classes. Silk is derived from the silkworm moth larvae, which feed on the leaves of mulberry trees; the worms were domesticated at least 3,000 years ago, and China enjoyed a monopoly on the production of silk until the middle of the sixth century AD, when two Persian monks returning from China smuggled some moth eggs and mulberry seeds into Byzantium. Although modern instruments such as humidity gauges have made it easier to control the environment in which silkworms are raised, the process has remained largely unchanged. The eggs of the silkworm moths are either exposed to harsh weather or dipped in brine to kill off the weaker specimens. Upon hatching, the larvae are fed constantly for about one month on mulberry leaves; during this time, they surround themselves with cocoons of a delicate fiber, itself produced by liquid secretions that harden on contact with the air. When the larvae finish spinning their cocoons, they are plunged into boiling water, which kills the new moth and loosens the fiber. Almost 3,000 feet of thread can be reeled from each cocoon.

Because of the reliance on land for cultivation, only marginal terrain was devoted to grazing, and very little grain was expended on animal feed. Animals that could be fed on household scraps, or that could scavenge for themselves, such as pigs and poultry, were the livestock most commonly raised for food. Water buffalo proved ideal for pulling plows in the warm, damp conditions of the south, but their ownership was beyond the means of most farmers. Limited numbers of horses and mules were used for transport; usually, it was cheaper for humans to haul or carry goods.

A similar situation prevailed in India, where a vegetarian diet was the norm for the vast majority of the population. Indeed, for adherents of Buddhism and Jainism in India, vegetarianism was not just a necessity but an expression of virtue: According to the doctrine of the transmigration of souls, even an insect might harbor what had been and could again become a human soul and should, therefore, not be killed. In India, however, more use was made of dairy products, especially ghee—boiled and clarified butter—which became an essential cooking medium. Vegetables or lentils were cooked in ghee, flavored with spices, and then diluted with coconut milk to make a curry sauce that turned a bowl of rice into a substantial meal.

The scarcity of meat did not restrict the potential variety of the Chinese diet. A poem written in the second century BC celebrated a banquet that included pigeon, goose, chicken, oriole, roast crane, magpie, bream, turtle, pickled pork, dog cooked in bitter herbs, five varieties of grain, and four kinds of wine. This was hardly everyday fare, but for centuries, the filling staples eaten by the ordinary Chinese have been enlivened by a combination of less copious dishes based on vegetables, eggs, or meat. The greater the profusion of the latter, the more luxurious the meal.

Traditions of local cuisine were established that closely reflected the geography and consequently the produce of each region. In the grasslands of Mongolia, wheat bread is staple, and meat, particularly lamb and mutton, is plentiful; dairy produce is also popular. In the time of the Tang dynasty, from the seventh to the tenth century, the taste for milk was widespread, but most Chinese today find it repulsive. The northern Chinese eat quantities of steamed bread, wheat noodles, and stuffed dumplings. Sichuanese cooking is characterized by the liberal use of chilies, which were introduced from Central America in the sixteenth century; in the Yangtze Delta, fish is important, and sweet flavors are popular. Ingredients are often cut into matchstick-size shreds and are usually cooked quickly at high temperatures—probably because of the shortage of firewood for fuel.

Far from being the parochial customs of isolated provinces, local traditions of agriculture and cuisine had by the fourteenth century become distinctive strands woven into a huge and productive society. Following the introduction of the fast-maturing Champa strain of rice from Vietnam in the eleventh century, which ripened in one-third less time than the native plant, the land could be made to yield two or three harvests each year. Yields were also improved by the careful preparation and conservation of soil. Transportation benefited from paved roads and the use of locks in a spreading network of canals, and a nationwide system of customs houses levied tariffs on goods as they were moved around. The growth of markets stimulated trade both within and among China's three main economic regions—the north, the lower Yangtze Valley, and the Sichuan basin—and opened up a number of specialized occupations to the rural peasantry, including forestry, papermaking, and weaving.

In other fields also, China's technological achievements during the period known in Europe as the Middle Ages were vastly superior to those developed in the West. Machines for reeling silk and spinning hemp speeded the manufacture of clothing. Doctors learned to diagnose diseases with a new accuracy. Coal was used in the smelting of iron ore; gunpowder revolutionized warfare; and mathematics and astronomy flourished. While visiting China at the end of the thirteenth century, the Venetian traveler Marco Polo was filled with awe at the bustling, teeming cities and the productive countryside. "Indeed it is scarcely possible to set down in writing the magnificence of this province," he declared; and of the city of Hangzhou, he wrote that "anyone seeing such a multitude would believe it a stark impossibility that food could be found to fill so many mouths."

For all governments of eastern and southern Asia, the problems of balancing the food supply with a growing population have been a pressing concern—not the least because a hungry populace can quickly become an army of rebellion. The shelves of the Han imperial library were filled with scrolls on agriculture and medicine as well as on astronomy, mathematics, and military matters. Very soon after the invention of printing in China in the ninth century AD, treatises containing advice on husbandry began to circulate widely. But only a tiny proportion of the population was able to read, and the spread of information could have only a limited effect. The most direct means of controlling production, as rulers soon realized, was through control of the land itself and the people who worked it.

Many forms of control were attempted. In northern China in the fourth century, for example, land was allotted to individuals for use during their lifetime, with provision for officials to receive larger portions according to their rank. But this system proved incompatible with the wet-field farming of the Yangtze Valley, as it failed to ensure that those who made the great investment of collaborative labor needed for irrigated farming to succeed were the ones who reaped the benefits. Other systems of land tenure varied from region to region. In some places, such as the mountainous regions of today's province of Zhejiang, peasants were usually independent; in others, they were virtually slaves, working under the close supervision of an overseer; in areas ravaged by warfare or underpopulated for other reasons, the government often took direct control. In general, the more productive the land, the more likely it was to be owned by landlords and worked by tenants.

Landlords clearly exercised immense control over the lives and fortunes of others, and the Song dynasty that ruled from AD 960 to 1279 developed a system of coopting members of this ruling class into the administration of the state. The wealthiest people in a village were required in turn to perform such duties as the collection of taxes, supervision of waterworks, arbitration of disputes, road maintenance, and when necessary, the distribution of famine relief. But even this system did not solve the perennial problems of government—for whenever the state devolved responsibilities upon local elites, they would build on their position of power to challenge imperial authority. Many took advantage of their status to evade their taxes, passing the burden on, through their managers and agents, to those in their sway.

From the fifteenth century onward, the administration of China was carried out in large measure by a class of educated scholar-gentry who owed their position to success in the civil service examination system rather than to hereditary wealth. The growth of this meritocracy, coupled with the development of a market economy in the countryside, led to a decline in the power of wealthy landowners—and the responsibilities of working the land increasingly fell to those most directly concerned, the peasant families who made up by far the majority of the population.

In Papua New Guinea, the heads of butchered pigs are impaled above stones that are being heated to cook their flesh. For certain peoples in New Guinea and the islands of the South Pacific, the pig is a holy animal that must be sacrificed to their ancestors, and once or twice in a generation, these peoples slaughter and eat the majority of their herds in a ritual feast. The roots of the practice may lie in the need to regulate the numbers of livestock so that they do not encroach on growing crops.

In some societies, animals deemed sacred are eaten; in others, their flesh is forbidden for the same reason. As the examples shown here and overleaf illustrate, attitudes toward food obey no universal pattern, often because they have evolved in response to very specific local conditions.

When early farmers discovered that a particular meat caused illness, they avoided it. When the raising of a certain animal conflicted with their essential way of life—pigs, for example, which yield no milk, are less vital to pastoralists than sheep or goats—they learned to do without its meat, however delicious. Over time, these customary habits frequently developed into religious prescriptions, which were themselves reinforced by other factors—the need to standardize behavior to maintain the social order, or to give expression to a people's sense of their unique identity.

THE LAWS OF DIET

The rural family household was the essential building block of the edifice that was the Chinese imperial state. The family was seen as reaching both backward and forward in time, encompassing both ancestors and unborn descendants; it was also linked by a network of kin groups, lineages and clans to other family units occupying a wide geographical area. In general, households were large, as the prosperity of each family depended on the number of able-bodied members it could put to work. On the other hand, because land was not inherited by the firstborn exclusively, farms were divided into increasingly smaller parcels, and most were less than five acres.

An eighteenth-century cotton wall hanging from northwest India depicts cows grazing and calves suckling in a paradisiacal landscape. The sacred status of cattle in Hinduism and the ban on eating their flesh probably derived from the value placed on living animals in agriculture. Cattle are still used for drawing plows in most of India, while their dung is used as a fertilizer and is the only cooking fuel for nine out of ten rural households.

The life of rural families was far from comfortable. In the sixteenth century, a local historian in Tancheng county in the province of Shandong, northeastern China, reported that although in a normal twelve-year cycle there were equal periods of abundance and scarcity, his own district was stricken by famine every year. When a new magistrate arrived in the county in 1670, the local residents reported that their land had long been "destitute and ravaged. For thirty years now, our fields have lain under floodwater or weeds; we still cannot bear to speak of the devastation." Floods and locust plagues had repeatedly destroyed their harvests, and thousands had been killed by gangs of bandits. "On top of all this came the famine of 1665; and after the earthquake of 1668, not a single ear of grain was harvested, over half the people were dying of starvation, their homes were all destroyed, and ten thousand men and women were crushed to death in the ruins."

In better times, the land yielded sufficient food, but not without continuous work. As soon as the winter snows had melted, the fields in Tancheng county in which winter wheat was not growing were turned and spread with animal and human fertilizer. In early May, this land was plowed and sown with millet and kaoliang; the earth was leveled with a wooden harrow and compacted either with a roller or by treading feet. As the crops grew, the fields were repeatedly weeded and the soil tamped down around the shoots. The winter wheat was harvested in June, and the land where it had grown was plowed and sown with soybeans. In August, the millet and kaoliang were harvested and carried to the threshing ground, and in October, the winter wheat was planted. Besides these staple crops, the farmers also grew vegetables such as turnips, cabbage, onions, celery, eggplant, and garlic.

From the sale of their produce, the farmers had to pay taxes to both the local and central governments. In addition to taxes on land and agricultural production, there were levies on sales of a range of items including salt, livestock, wine, and cotton goods, and the government could buy produce from farmers at below-cost prices. All adult males were liable to be drafted for corvée labor—to build dikes, irrigation

A fifteenth-century Italian illustration from a Hebrew law code shows the ritual slaughter of fowl and oxen. While many of the Jewish dietary laws laid down in the book of Leviticus may have evolved because they were hygienically beneficial, others have more complex origins. The pig, for example, was seen by early Middle Eastern pastoralists as a very different kind of animal from the sheep, goats, cattle, and camels that constituted their herds; and it was possibly because the pig did not fit the perceived norm that its flesh was outlawed.

canals, and other public works—or else had to pay an extra tax in lieu of labor. Many households were required to furnish a militiaman in times of emergency.

Unsophisticated though it may seem in comparison with modern Western agriculture, which made increasing use of technology and modern science from the eighteenth century onward, it was this labor-intensive system based on the tilling of small plots by family units that enabled China to support its growing population into the twentieth century. The very size of the work force was one factor that encouraged such conservatism: Where labor was cheap and plentiful, the introduction of machinery that saved on man power was unnecessary, and indeed would have created new social problems by throwing thousands if not millions out of work.

Nevertheless, the growing population did place increasing pressure on agricultural resources. In 1740, the emperor noted that "the population is constantly increasing, while the land does not become any more extensive," and he urged his subjects to cultivate every available plot of ground "at the tops of the mountains or at the corners of the land." At the end of the first millennium, China's population was around 100 million; by 1850, this number had risen to 450 million. In the twentieth century, it became apparent that the needs of the population could no longer be met without radical social reforms and technological development.

These became the urgent priorities of the Communist government after the establishment of the People's Republic of China in 1949, and they were acted upon with a speed and ruthlessness that turned Chinese society upside down. Land reforms completed in 1952 confiscated the holdings of landlords—two million of whom were killed in the process—and redistributed their land to poor peasants. There then began a program of collectivization, modeled on the experience of the Soviet Union in the 1920s, which involved the elimination of private ownership of land. By 1957, it is estimated that 96 percent of the rural population was incorporated into agricultural cooperatives in which all equipment, animals, and labor were pooled for the common good. The cooperatives were in turn organized into self-sufficient administrative units known as communes, each comprising an average of 5,000 households.

Such drastic changes could never have taken place had not the Chinese been accustomed for centuries to a hierarchical society in which obedience to one's elders and those in authority was not just a social virtue but a way of life. But in their attempts to substitute the state for the family as the focus of loyalty for all Chinese, the government created unexpected disturbances. Reverence for the past had for centuries been a mainstay of social stability: Evident in the chronicles of past dynasties painstakingly compiled by court historians and in the works of artists and writers who adhered to ancient traditions, this reverence's most basic expression was the ancestor worship of individual families. By seeking to eradicate all past models, the Communist leaders ensured that the birth of the new order was accompanied by confusion and disorientation.

In addition, the government's policies were often based on information that did not accord with reality. Following the establishment of the rural communes, the government came to believe, in line with inflated production figures supplied by ambitious commune leaders, that China had a massive surplus of grain, and it ordered a reduction in the area of land given over to rice growing. In fact, during the years between 1958 and 1961, China was suffering probably the worst famine in its history, causing the deaths of as many as 30 million people.

Subsequent reforms, based more on a realistic assessment of China's needs than on ideological rigor, improved the lot of the long-suffering peasantry. Between 1965 and 1975, it is claimed that the area of irrigated farmland was increased by one-third. In the 1980s, individual peasant households again became the basic unit of production, and they were permitted to sell surplus produce at rural markets. Overall, the food supply per person rose after the Communists came to power.

At the beginning of the 1980s, the government instituted a program designed to reduce population growth. Families were officially forbidden to have more than one child and lost welfare benefits if they exceeded this limit. Although the program represented a degree of state interference in private life that would have been unacceptable in the Western world, it was the logical response to the inescapable fact that China's arable land—only 11 percent of its total territory—could not support an ever-growing population.

Other countries around the world looked to China with keen interest, for in the 1980s there was a growing awareness that the problems of food supply were present

In this sixteenth-century Japanese garden—a frozen encounter between resistant nature and the human urge to arrange and transform—the forms of mountains and rivers are tamed and reflected in a cultural mirror. Introduced from China in the seventh century, the art of creating these austere and compressed landscapes accorded well with the nature veneration of the ancient Japanese Shinto religion. After Zen Buddhism came to Japan in the early thirteenth century, again from China, such gardens were created for Buddhist temples as ideal settings to foster meditation.

on a global scale. In 1990, it was estimated that 700 million people in the Southern Hemisphere were without adequate food—or in other words, for every affluent person in the north, another was close to starvation in the south.

Apart from limiting the size of families—and in most countries of Asia and Africa as well as in rural districts of China itself, effective birth-control programs have so far proved almost impossible to implement—only the application of advanced scientific procedures to crop growing could alleviate the scale of world hunger. In the nineteenth century, the Western world solved its own food crisis by the mechanization of agriculture and the development of refrigeration, canning, and improved transportation. In the late twentieth century, biotechnology—the range of disciplines associated with the genetic manipulation of living organisms—appeared to offer hope of similar success. For example, crops that have built-in resistance to pests—and therefore do not require the use of chemical pesticides—could greatly reduce the loss in production caused by weeds and disease, which at present destroy as much as 40 percent of the world's crops. But all such developments are vulnerable to political and economic pressures. One variety of hybrid rice now grown in over one-third of China's approximately 80 million acres of rice paddies has increased harvests by 25 percent—but, determined to profit from this advance, China sold rights for the exclusive sale of the seed to two American companies, thereby limiting the number of poorer countries that could benefit.

Some Chinese hybrid seeds have proved unsuitable for cultivation in Western countries because they require so much labor. The plants made few seeds, and without a large and cheap work force to fertilize them by hand, they offered scant profits. But this very obstacle demonstrates that the agriculture of China and other Asian countries, often viewed by outsiders in the past as backward and inefficient, has become a model of increasing relevance to other countries whose massive and growing populations appear to handicap all plans for a sustainable future. Labor-intensive, sparing of natural resources, and with a strong emphasis on grains and vegetables, the agriculture of Asia is greatly productive in terms of yield per unit of land. Not the least of its exemplary characteristics has been the degree of communal effort necessary for survival.

MINING THE EARTH'S RICHES

In Africa, India, and North America, there are gold mines that extend so deep into the earth's crust their walls are hot. Without air conditioning, the miners would die. When they return to the surface, in elevators that travel slowly so that their ears do not pop, their faces are unnaturally pale.

Humans are not built to burrow under the surface of the planet, but from the time they first began to fashion sharp tools and bright ornaments, they have devised methods of extracting the riches embedded in the earth. Flint was quarried to make blades beginning around 40,000 BC. Copper was mined by the ancient Egyptians, and by Roman times, the metal ores extracted from both surface and underground sources included iron, gold, silver, lead, tin, and zinc. The Industrial Revolution accelerated demand for coal and, later, for other fossil fuels such as petroleum and natural gas, all formed by the decomposition of prehistoric organisms.

The rate of material progress generated by the earth's minerals has been matched by the increasing impact of mining on the environment. The most conspicuous effects are those of surface mines—both those worked by manual labor, as shown in the photograph on the left of a gold mine in Brazil in the 1980s, and those exploited by mechanical excavators, the largest of which can remove 6.3 million cubic feet of coal per day. But the impact of underground mines is no less severe: They can cause subsidence, contaminate water sources, and despoil the surrounding land. The furnaces for smelting silver ore from Roman mines in Spain, for example, consumed 500 million trees, clearing an area of almost 7,000 square miles. The safe disposal of radioactive waste from the mining and processing of uranium and plutonium for nuclear power now challenges the technology that, as shown on the following pages, has enabled humans to penetrate ever farther from their given habitat.

A clay tablet of the sixth century BC (right) depicts Greek miners quarrying for clay. A ladder is propped against the side of the pit; the hanging amphora probably contained refreshment. On the far right, a fifteenth-century German woodcut shows a natural seepage of petroleum being collected in a jug. Oil was believed to cure many ailments: "He who pants with a cold or cough of long standing, let him rub the chest with it, and it helps," declared a contemporary pamphlet.

In tunnels dug beneath the Japanese island of Sado, where gold was discovered around 1600, miners in a nineteenth-century woodcut by Hiroshige work by the light of burning wicks in vessels of oil.

OPENING THE SEAMS

Although most mines were dug and worked by muscle power until the nineteenth century, the impact of picks and shovels was by no means negligible. In Austria during the second millennium BC, there were as many as 500 copper mines with timber-lined underground tunnels in the region around present-day Salzburg, while flint was mined in England from galleries supported by pillars of unworked rock.

In addition to human and animal muscles, the most long-established and enduring source of energy harnessed for surface mines was water, which was employed by the Romans to break up banks of loose soil. Beginning in 1853, the use of high-pressure waterjets during the California gold rush altered the contours of the land, shaving soil from the hillsides and building up new hills out of the debris.

A German watercolor dating from 1480 depicts an underground mayhem of pick-wielding miners. Because of the value of the iron ore they extracted, German miners were among the most privileged of medieval workers: Many were exempt from taxes and military service.

In this silver mine in Lorraine, depicted in a sixteenth-century German engraving, picks, hammers, and wedges remain the miners' essential tools, but the removal of mined ore has been streamlined by the use of manual winches and wheeled carts in vertical shafts and level, horizontal tunnels.

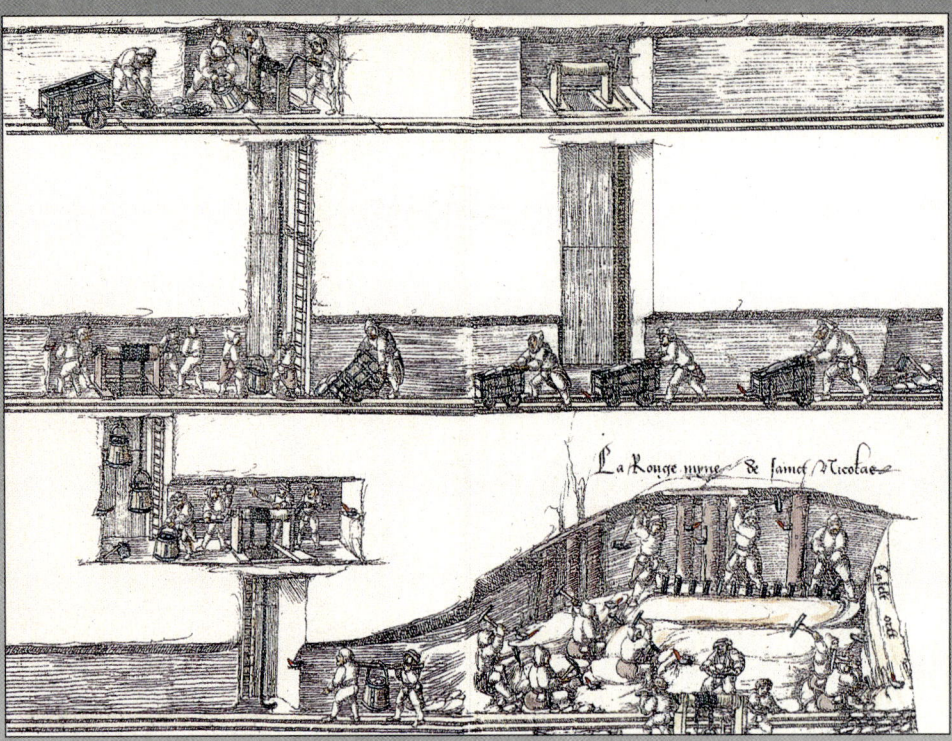

Sixteenth-century woodcuts from a treatise by German mineralogist Georgius Agricola show drainage and ventilation: Above, a man works the piston of a suction pump to raise water; above right, two men turn cranks to send air underground.

A steam engine *(right)* at an early-nineteenth-century pithead in England, feeding off the hemispherical boiler under the smokestack, powers the gear that raised and lowered both miners and coal. On the right, a loaded cart stands on a weighbridge.

INCREASING THE YIELD

In medieval Europe, machines powered by humans, animals, or streams were devised to bail water from flooded shafts and raise minerals from ever-lengthening tunnels. To mine iron ore—used for tools, weapons, and armor—hammers and wedges were supplemented by gunpowder and by fires set to fracture a rockface. Dwindling supplies of wood for the smelting of ores led to an increased use of coal, especially in England and Wales, whose output by 1700 amounted to five times that of the rest of the world. The development of steam power by the Scottish engineer James Watt around 1765 provided a new source of energy with which to pump out floodwater, and after the introduction of steam-driven rotary and piston drills in the early nineteenth century, to dig ever deeper.

A seventeenth-century Chinese woodblock print shows workers binding bamboo pipes that will be used to carry brine from underground salt mines. The pipes were also used as conduits for natural gas, which was tapped from salt mines as early as the fourth century. This gas was burned to evaporate brine and also to provide heating and lighting for nearby buildings; in the seventeenth century, some entire towns were illuminated by its flames.

APPROACHING THE LIMITS

During the last century, a twentyfold increase in demand for fossil fuels was met by rapid advances in both the techniques and the scale of mining. The first oil well was drilled in Pennsylvania in 1859; by the end of the century, oil was being extracted from every continent except Antarctica. Natural gas that was trapped in strata under the land or sea became the world's single largest source of energy by 1970. The first nuclear power station, offering an alternative to the burning of fossil fuels, was opened in Britain in 1956.

By the late 1980s, methods were being developed to recover oil untapped by earlier technology, including the use of chemicals that generate heat, radio frequency waves that make heavy oil less viscous, and the fracturing of impermeable rock by fluids pumped underground at high pressure.

A forest of wooden derricks *(left)* covers a Pennsylvania hillside in 1865. Demand for oil—first used for lighting—increased after the development of the internal-combustion engine in the 1880s.

Five tugboats *(right)* tow a production platform from Stavanger in Norway to its station northeast of the Shetland Islands, where it began drilling for both oil and gas in 1975. Drill shafts and conductor pipes are housed in the concrete supports; undersea storage compartments have a capacity of 1.1 million barrels of oil.

A worker at the United Kingdom's center for the reprocessing of spent nuclear fuel *(right)* watches on a screen the operations he directs by means of the remote-control manipulator in his hand. The plant extracts reusable uranium and plutonium and converts high-level radioactive wastes into solid glass blocks, which are sealed in stainless steel containers.

THE PLOW AND THE CROSS

Hollow-cheeked peasants, their bellies bloated from hunger, wandered aimlessly through the marketplace of Tournus in eastern France. Even if they had come with well-filled purses, or a basket of goods to barter, there was nothing to buy: The land lay in the grip of the worst famine that anyone could remember. Throughout the centuries of privation that followed the fall of the Roman Empire, life was never easy for people striving, with poor tools and ill-fed animals, to scratch a living from the soil, but the year AD 1030 brought conditions that were nothing less than catastrophic. The weather, all over Europe and far to the east, turned cruel. For three years in succession, the land was battered by violent, drenching storms. Seeds sown in the sodden ground were washed away; harvesttime brought only a crop of weeds.

The contemporary Burgundian monk Radulfus Glaber painted a grim picture: "After men had eaten beasts and birds, under the pressure of rampant famine, they began to eat carrion and things too horrible to mention. . . . It is terrible to relate the evils which then befell mankind. Alas," he said, "a thing formerly little heard of happened: Ravening hunger drove men to devour human flesh! Travelers were set upon . . . and their dismembered flesh was cooked over fires and eaten."

Yet at the end of the year 1033, according to Radulfus, such suffering was miraculously ended. "At the millennium of the Lord's Passion, the rains of the thunderclouds ceased in obedience to divine goodness and mercy. The sky began to smile." Famine, he assured his readers, was a thing of the past, for the earth was again fruitful.

This transition from dearth to fecundity was not as instantaneous or as universal as Radulfus described it. But his statement was in essence true: The time and place in which he lived did indeed witness significant changes not only in the weather but also in the whole web of relationships between humankind and the natural world. From the eleventh century until the beginning of the fourteenth, the men and women who worked the land enjoyed the benefits of a kinder climate, and the population rose at a rate far greater than that achieved at any time in the previous 1,000 years. The pressure of extra mouths to feed spurred improvements in the techniques and technologies of farming, increasing the area under cultivation. And as forests were put to the axe and soil to the plow, Europeans began to perceive that the world was more malleable and its boundaries less finite than they had been taught to believe.

These developments were not simply a set of responses to environmental factors outside human control; they were intimately bound up with the social institutions that dominated people's lives and molded their perceptions. No medieval European was beyond the influence of the rules of conduct enjoined by the Church and its teachings on the relations between human beings and the rest of God's creation. No lord or peasant lived outside systems of landownership and political authority that, while differing from one part of Europe to another, invariably dictated how labor would be

---

Laborers load a cart with newly cut logs in this detail from a fifteenth-century Flemish tapestry; behind them fellow woodchoppers hew and saw, leaving raw stumps where once was woodland. In addition to being the only household fuel, wood was essential to every medieval industry: It fed iron-smelting and glassblowing furnaces, was shaped into wine casks and vats, and was used to construct carts, ships, bridges, textile machinery, and most houses. In the top right corner is one of Europe's newly expanding towns, in which the building of an average house consumed twelve mature trees.

performed and wealth distributed. But such institutions were not inviolate, and as the fruits of affluence began to permeate society, the hold of these systems was loosened and a new order began to emerge. By the end of the fifteenth century, an outward-looking, intellectually vibrant continent was poised on the brink of an era of expansion that would spread European approaches to nature across the globe.

Europe's weather was subject to broad variation. The climate of a marsh on the North Sea bore little resemblance to those of an Alpine meadow or the Mediterranean plains. But in the north and west of the Continent especially, the eleventh century saw the arrival of markedly warmer and drier conditions, with milder winters and longer growing seasons. In some areas, rainfall diminished sufficiently to cause a drought, and settlements primarily dependent on rain for water had to be abandoned; elsewhere, particularly in highland regions, villages that had been deserted for 1,000 years were now repopulated, thanks to an increasingly hospitable climate.

Travel became easier. In the Alps, where glaciers began to retreat and the snow line rose higher, passes became accessible to travelers for more of the year. Storms subsided in the North and Baltic seas and the Atlantic Ocean, allowing Norse seafarers to set out on voyages of exploration and conquest. They ventured beyond the known world to Iceland, Greenland, and the easternmost tip of North America; they penetrated Russia and northwestern France through the European rivers; they sailed through the Strait of Gibraltar and even made an appearance in Byzantium.

The Norse did not take to their longboats simply through wanderlust. Like other Europeans, they were feeling the pressure of an increasing population. Some 90 percent of Europe's people lived in small rural communities; these had once been self-sufficient, but now, for the first time in centuries, there were more mouths to feed than the local fields could sustain. Meanwhile, Europe's towns also expanded, attracting artisans and merchants who did not work the soil themselves, but who still had to eat. Townspeople made greater demands on the countryside than did the peasants, consuming 15 to 25 percent of the food produced in the hinterland.

Responding to these demands, farmers who had fully exploited the best lands in their home territories were now forced to cultivate less-promising ground or to venture into altogether unknown country. In north and south alike, settlers moved tentatively into frontier lands; in mountainous regions from the Alps to the Pyrenees, shepherds guided their flocks across the high passes in search of new pastures.

Every part of the Continent had its own variations in landscape, soil, and climate

This map was made in the 1440s for Edmund Rede, the lord of Boarstall Manor, whose ancestor is shown in the lower field receiving title to the land from Edward I. Isolated and self-sufficient, Boarstall in Buckinghamshire, England, was a typical medieval village centered on its manor house and church. Crops grown in the common fields—which were cultivated on a three-year rotation cycle of winter wheat, spring oats, and a year fallow—provided the main part of the inhabitants' diet. For meat, the villagers resorted to rearing pigs and hunting in the adjacent woodland. The illustrations from a fourteenth-century Psalter below show a man catching partridges in a drop net and women using a ferret to drive a rabbit from its hole into a trap.

and posed its own problems. Low-lying lands, from the East Anglian fens and the coast of Flanders to the marshes of the Italian Veneto, had to be drained before seed could be sown. In the arid expanses of southern Spain, where the Arab and Christian worlds met, settlers had to dig irrigation canals to make the land bear fruit. In the forested north, German migrants pushed east into the Slavic lands, hacking through vast woodlands to make way for the cultivation of grain while simultaneously striving to sow seeds of Christian faith among the indigenous pagan population, which was still dedicated to the worship of forest-dwelling gods.

The lords who owned land in these frontier regions swiftly realized that an influx of willing colonists, with strong backs and healthy appetites, could increase the value of their holdings. A Slavic chronicle told how the perspicacious count Adolf of Schauenberg brought new life to the underpopulated territories lying northeast of the lower reaches of the Elbe River. Making much of the land's unfulfilled potential, the count "sent messengers into all parts, namely to Flanders, Holland, Utrecht, Westphalia, and Frisia, proclaiming that those who were in straits for lack of fields should come with their families and receive a very good land—a spacious land, rich in crops, abounding in fish and flesh and exceeding good pasturage."

A further incentive to immigrants was the promise of far greater freedom and social mobility than their stay-at-home counterparts enjoyed. For the peasantry, especially in northern Europe, dwelled at the bottom of the rigid hierarchy of property and power relationships known as feudalism. This system—dominant in Germany, Flanders, England, and northern and central France, but also present in Catalonia, Sicily, and elsewhere in the south—was the product of centuries of political chaos, tribal invasions, and social flux that had followed the collapse of the western portion of the Roman Empire. It was rooted above all in a social contract between those who were—in both military and economic terms—the strong and the weak.

At the top of the social pyramid stood great warlords with large estates. Since the introduction of the stirrup in the eighth century, horses had played an increasing part in warfare, and battlefields had become the scene of clashes between charging, heavily armed and armored adversaries on thundering steeds. Only those individuals affluent enough to purchase and feed a horse—and to fit themselves out with the appropriate apparatus—could enter the military elite. And almost without exception, the basis of their income was land and any surplus wealth it yielded.

The warlords parceled out their properties to fighting men of lesser fortunes in return for their allegiance and support in times of strife. This emergent class of aristocratic warriors bestowed their protection on the mass of humble noncombatants; in exchange, these peasant dependents grew the crops that sustained their masters' military adventures. By the tenth century, the toilers in the field were not simply the social subordinates of the landed elite but their legal inferiors as well, with their second-class status fixed by custom and enshrined in law. They had become manorial serfs and were tied to the land they worked. They might possess a small patch of ground and some rudimentary tools, and perhaps also a pig or some fowl, but their daily and seasonal occupations, dwelling places, marriage plans, and domestic arrangements were subject to the consent of the lord of the manor.

Outside the feudal lands, peasants—in theory—enjoyed greater freedom. In Spain, for instance, the territories of León and Castile, newly recaptured by Christian forces from the Moors, attracted Basque, Galician, and Asturian colonists. Though dominated at times by local grandees, these landholders kept their independence and never sank to the status of serfs. In northern Italy, the citizens of the burgeoning towns owned much of the property in the surrounding countryside and gradually took control of the agricultural life within their own districts, setting quotas for grain, controlling cereal prices, and banning or encouraging food exports as appropriate in periods of dearth or glut. In Pisa, the authorities directed the local peasantry to plant a prescribed number of fruit trees annually and required each rural household to cultivate beans and a cabbage patch. Their official counterparts in Parma ruled that all fields must be plowed four times before any crop was sown, while the town of Modena designated the amount of land that was to be allocated to vines and set an annual date for the harvesting of grapes. Yet even among a free peasantry, autonomy was discouraged: The city of Milan passed a statute at the beginning of the twelfth century requiring that peasants "show reverence to their lord."

The all-powerful institution that supported this social order and laid down the rules of conduct for its members was the Christian church. On all but the farthest fringes of the Continent—Muslim Spain, the lands of the Balts and Slavs, or the forests and fjords of Scandinavia—the hegemony of Christianity was complete. And as well as being a source of moral guidance and a dispenser of the sacraments that promised comfort in this world and salvation in the next, the Church was the wealthiest landowner in Europe. Its coffers bulged with rents, fees, tithes, bequests, and charitable contributions. Its jeweled reliquaries, embroidered vestments, and illuminated Gospels no longer provided easy pickings for heathen marauders: The raiders themselves, or their descendants, had transformed themselves into baptized members of the congregation. And its wealth continued to grow as princes, for the good of their souls, made gifts of vast estates with fat rent rolls to their bishops, and the lords of rural manors donated land or subsidized the building of new religious houses.

Such offerings made sense in material as well as spiritual terms: Any district stood

In a fourteenth-century Flemish manuscript illustration, a man carries a sack of grain toward a mill, where it will be ground into flour. Windmills of this type, which were based on earlier Arab and Roman models, first appeared in the northern part of Europe in the late twelfth century. The boxlike body, mounted on a sturdy oak post, could be turned into the wind by pulling on the tail pole protruding from its base.

Miniatures from a breviary produced in Flanders in the late fifteenth century depict a range of agricultural tasks associated in northern Europe with particular months of the year. These include, from left to right: plowing, spreading the land with animal manure, and sowing in March; sheepshearing and reaping the grain with sickles in July; knocking acorns to the ground to fatten hogs in November, after which the animals will be slaughtered and their meat salted to preserve it through the winter; and spinning wool indoors in February while sheep huddle for warmth in the fold. In three of the four pictures, the unchanging cycle of human activities is overseen and sanctified by the ever-present church.

to benefit by the presence of a thriving monastery in its midst. And from the eleventh century, the monasteries themselves were becoming a new type of institution, with consequences that were to affect the natural world. Except within the still-pagan peripheries, the traditional task of making converts was no longer necessary; at the same time, the established monastic orders seemed to have lost the spiritual purity and rigor of their early Christian precursors. New members of the orders turned their attention to the desert fathers who, in the first centuries of the Christian Era, had withdrawn into the wilderness, seeking salvation and enlightenment through physical labor, prayer, and meditation. Alone or in small bands, they set out to establish hermitages or small communities in wild, unpeopled places. By the beginning of the twelfth century, this revised approach to the cloistered life had become a significant force within the Church, and two new orders, the Carthusians and the Cistercians, had set up the first of many monasteries in remote regions: the former in the mountainous Dauphiné in southeastern France, the latter in Burgundy's Saône Valley, with its dense forests and unpromising, sandy soil.

Pulling up dead and tangled roots, cutting down trees, clearing the ground of

stones, coaxing reluctant earth to receive the plow, the members of the new ascetic orders gradually transformed the landscapes to which they had retreated. The Cistercian monk Bernard, later to be canonized, spoke with pardonable pride of the work done by members of his own community at the mountain abbey of Clairvaux. Looking up from the cloister in the valley, Bernard could see two hillsides: one slope made golden by the ripening grain that his brothers had planted, the other marked off with orderly rows of vines. "Each of them," observed Bernard, "offers to the eye a beautiful sight and supplies a needful support for the inmates."

On the wooded tops of these hills, monks gathered dry branches for fuel and cleared the ground to allow the native oak, lime, and beech trees to grow without interference. Down below, closer to the monastery itself, the Aube River ran through the valley on a course that the monks had diverted to prevent dangerous flooding, to irrigate their fields, and to power a series of mills. On its banks stood young orchards, heavy with blossom and promising rich yields of fruit. To Bernard, this carefully manipulated landscape was more than a tribute to the monks' good husbandry—it was a living sermon full of spiritual truth. "It greatly soothes weary minds, relieves

Women afflicted with epilepsy or chorea are half-carried, half-dragged on a pilgrimage to a local church in this seventeenth-century copy of a lost original drawing by the sixteenth-century Flemish artist Pieter Brueghel the Elder. It was believed that dancing to bagpipes would rid victims of seizures for twelve months; one woman is held fast while a piper plays directly into her ear. Disorders of the mind and body abounded in the Middle Ages, largely attributable to inadequate nutrition and to bread made from contaminated grain. Bread infected with the ergot fungus spread a fatal disease known as Saint Anthony's fire, which caused burning pains, hallucinations, and gangrene.

anxieties and cares, helps souls who seek the Lord greatly to devotion. . . . While I am charmed without by the sweet influence of the beauty of the country, I have not less delight within on reflecting on the mysteries that are hidden beneath it."

If the transformation of the wilderness was a spiritual mission for the brothers of Clairvaux and like-minded communities, it was also a hardheaded agricultural enterprise. In Flanders, for instance, the abbot Guillaume de Ryckel, recruiting peasants to help clear and work new lands attached to the monastery of Saint-Trond, set out the ground rules for his prospective tenants. For the first six years of their tenure, new settlers would be free to plant whatever crops they chose, in any quantity or sequence. From the seventh year, once the farms were established, the tenants were told to "observe the common custom of sowing, so that in one year they sow wheat or rye; in the second barley or oats or whatever spring crop is usually grown; and in the third year they will sow nothing." The abbot's prescription reflected contemporary wisdom. To keep the ground from becoming exhausted, farmers understood the need to rotate their crops, to alter the crops grown, and to allow fields to lie fallow.

In southern Europe, the pattern of cultivation had not changed significantly since Roman times. Peasants worked small, enclosed fields, using a light plow that was suitable for drier, more friable soil. Farmers in the Mediterranean lands could not expect enough summer rain to give crops that were planted in the spring any reasonable chance of survival, so the southern practice was usually to sow only one cereal crop in a year. Nor were hot, dry summers the only impediment to high productivity. Lower yields of grain meant less fodder to sustain livestock, and fewer farm animals meant less manure to fertilize the fields—the cycle was self-perpetuating. Nevertheless, in compensation for its relative paucity of cereal crops, the south had other riches: olives for oil, plentiful grapevines, and a long growing season for a cornucopia of vegetables and fruits that few northerners had tasted.

North of a line that stretched approximately from the Alps to the Loire Valley lay a colder, wetter region, scene of a slow but significant agricultural revolution that had begun during the eighth century and reached its peak in the eleventh and twelfth centuries. During this period, the farmers of the north adopted a pattern known as the three-field system. They divided their land into three parts: a section for winter sowing of wheat or rye; a second field reserved for spring crops such as oats, beans, and barley; and a third piece of ground to lie fallow. Every year, the functions of each field were rotated. The result was a virtual doubling of yields.

Whereas a light wooden plow had for centuries been adequate for farming the chalky lands of northern France or southern Britain, the heavy clay soils of recently cleared valleys in the north required the use of a heavy, wheeled iron plow of a type used in the Slavic lands that turned up the earth in furrows. The old light plows had needed only a single draft animal to pull them—a donkey, mule, ox, or in times of desperation, a human being. The weighty newcomers required the combined

A richly robed surgeon directs the activity of his pharmacy in this miniature from a fifteenth-century French manuscript. The assistant on the left grinds a medicine prepared from the ingredients in the jars behind him, while the youth on the right collects herbs from an enclosed garden. Beyond the arch lies the examination room. Qualified surgeons and physicians were capable of setting limbs, cauterizing and dressing wounds, and performing operations for hernia, gallstones, Caesarean section, and plastic surgery, during which anesthetics of mandrake or opiate-soaked sponges might be used. But they had little to offer sufferers of such incomprehensible diseases as malaria, scurvy, diphtheria, or typhoid.

strength of six to eight oxen. Such resources were beyond the means of a single peasant and compelled farmers to pool their equipment, thus reinforcing the collectivist, communal nature of village settlements. In addition, these plows changed the face of the northern landscape. Because maneuvering a hitched-up team within the confined space and awkward corners of a small, rectangular plot proved a frustrating and time-consuming business, there evolved by the twelfth century a patchwork of large open fields, subdivided into long, narrow strips for more efficient plowing. An ox team could cover considerable ground before it had to change direction.

In general, few animals were kept on farms. Except in the most mountainous regions, meat and dairy products formed only a small part of the diet for the bulk of the population, and most of this meat derived from the hunting of wild animals in nearby woodland. The only creatures reared exclusively for meat were pigs, and these were virtually self-sufficient, being allowed to forage among the oaks and beech trees in the woods. Sheep were raised for their wool; cattle were important for their hides and for milk or meat; horses, donkeys, mules, and oxen earned their keep as means of transportation or as beasts of burden. These animals were worked hard, but enlightened farmers recognized them as valuable resources and treated them as well as their modest means permitted. "You must keep your plow beasts so that they have enough food to do their work, and that they be not too much overwrought when they come from the plow," advised Walter of Henley, bailiff in charge of the manors farmed by an English community of Dominican monks, "for you shall be put to too great an expense to replace them; besides, your tillage shall be behindhand."

Walter took pride in the smooth running of his farms and felt he had learned enough to write a text on husbandry that would help other thirteenth-century agriculturalists. The farms he managed grew wheat for bread, animal feed—probably the oats that in times of famine would nourish human beings as well—and barley, a difficult crop but one deemed worth the trouble, for the sake of the beer and ale that could be brewed from it. He might decide to devote some land to inedible crops for industrial purposes, such as hemp for ropemaking or flax for cloth.

The warm, dry summers in this era of mild climate might even have tempted Walter to plant a vineyard: Grapes were now grown successfully in his own southern English countryside and on the Continent as far north as Flanders. But the heartland of the European wine trade lay to the east and south. Beside the well-trafficked rivers of central France, Burgundy, Lorraine, the Rhineland, and the Moselle Valley, large quantities of wine were produced for export, a trade stimulated by the sacrificial role of wine in the Mass. In 1245, the Italian monk Fra Salimbene, visiting the vineyards of Auxerre, observed with a tinge of envy that "these people sow nothing, reap nothing, and gather nothing into their barns. They only need to send their wines to Paris on the nearby river, which goes straight there. The sale of wine in this city brings them a good profit, which pays entirely for their food and clothing."

In the fourteenth century, the proliferating vineyards of Bordeaux, ideally placed for the cross-channel and Mediterranean maritime routes, became a powerful presence in the trade. The wine pressed from the grapes of the south made a fair exchange for the grain surpluses of the north, and both parties in these transactions prospered.

"Nose to the ground, I crawl forward," ran a riddle of the times. "My nose is the gray enemy of the forest." To countryfolk, the answer was obvious: the iron tip of the plow. And as agricultural life began to move to a more insistent rhythm, other simple but significant improvements took their place in the lives of Europe's farmers.

Wheeled implements in use by the thirteenth century included the spinning wheel, for turning flax or wool into yarn, and the wheelbarrow. Mills powered by flowing streams—used in the Middle East and Egypt since at least the first century BC—proliferated across the landscape. In the province of Picardy on the northeastern edge of France, manorial records noted that 40 mills were built between 850 and 1080, and more than 200 in the next 120 years. In England, the *Domesday Book*—the great statistical survey compiled by William the Conqueror—counted some 6,000 mills in operation at the end of the eleventh century. These devices could be turned to many uses: grinding grain for brewing or baking; pounding and processing cloth; or powering the hammer and bellows for a blacksmith's forge. And in those regions that did not have fast streams, the vanes of windmills rotated high above the flat terrain.

Following the development of a new form of collar harness with long leather traces and a flexible breast strap, horses came into their own as draft animals. A pair of horses harnessed in tandem in the new manner could pull three times as much as a team controlled with the traditional yoke, which tended to choke any beast pulling hard on a heavy load. The horse became an even more efficient helper after the introduction of nailed metal horseshoes, which provided better traction on difficult terrain and allowed horses to work longer and pull harder, with less fatigue. A horse, which ate primarily grain, was more expensive to keep than an ox, which fed on cheap hay. But in northern Europe, with its relatively productive three-field system, the horse's increased efficiency outweighed its greater maintenance costs; in the south, where grain was scarcer, a single ox or mule still pulled the lighter plows.

The blacksmith, shoeing horses and mending plows, became an indispensable member of every rural community, and the metals he worked were no longer luxuries or rarities in regions with no mines of their own. River barges and small seagoing ships were still the cheapest form of transport—and would be until the advent of the railroad in the nineteenth century—but an increasing amount of goods was carried by carts along improved roads whose reach was extended by a proliferation of bridges. Along their cobbled paving traveled silver from the Harz Mountains, iron from the Slavic lands, and tin from Cornwall to markets all over the Continent.

The ring of metal axes resounded everywhere, as peasants chopped down the woodlands. The forest had for centuries been used by peasants as a grazing place for their swine and as a source of domestic fuel. Their customary rights were enshrined in longstanding agreements with the lords of their manors: In exchange for the right to gather deadwood, peasants might be expected to weave a cloth garment every year for their landlord; or as a fee for letting their sows run free among the lord's oaks and beeches, they might owe a piglet from the animal's next litter. But the modest use of the forest for subsistence now yielded to a more thoroughgoing exploitation. Business-minded rustics collected wood far beyond their needs and sold the surplus to residents of the expanding towns. There was also modest profit to be gained by the gathering of squirrel skins, resin, charcoal, bark, wax, and lime for a variety of crafts and

ECHOES OF A PAGAN PAST

Two centuries after Christianity replaced paganism as the official religion of western Europe in AD 391, Saint Martin of Braga complained that the peasants were still participating in animist rites, such as burning candles before rocks and plants and making offerings of bread to hallowed streams. His denunciations had little effect on their behavior: Deep-rooted beliefs in good and evil spirits, and in the need to propitiate the unpredictable forces of nature through pagan rituals, persisted throughout Europe at least until the beginning of the Reformation in the sixteenth century.

Sometimes the old was subsumed by the new: Legendary tales of folk heroes became identified with Christian saints, and certain ritual objects—from mistletoe to eggs—were endowed with a Christian significance. But many customs escaped altogether the net of Christian doctrine. The human-shape mandrake root, for instance, shown in a fifteenth-century illustration on the left, was believed to utter a shriek that was fatal to any human listener when it was pulled from the ground; therefore, a dog would be tied to the root in order to pull it up. Similar beliefs are illustrated on the following pages, some of them the source of surviving superstitions.

industrial processes. But the greatest demand was for timber—to construct new houses for the swelling population, to build merchant shipping, and to make the barrels and chests that would carry the goods. For the first time, forests had to be managed, with selective timber cutting and the tending of replacement seedlings.

As the trees fell, old and new interests collided. The lords of the manors employed gamekeepers and bailiffs to prevent poaching. Legal documents bristled with disputes over rights of access and exploitation: lords against their neighbors, peasants against each other, indigenous farmers against incoming monks, one abbey against another.

The expansion of land under cultivation also led to conflict between farmers and pastoralists—the people whose livelihoods came from the raising of sheep or cattle, and whose open grazing grounds were now being eroded. Shepherds and cattle breeders in Iberia, the south of France, and Italy were forced to move permanently to new pastures or to undertake the seasonal migrations known as transhumance—the twice-yearly movement of livestock between mountain and lowland pastures, allowing intensively grazed ground time to sprout new grass before the herds or flocks returned. Even so, as the cloth trade burgeoned throughout Europe, the peacefully grazing flocks became themselves part of a vast commercial and industrial mechanism: If the farmers of one district drove out the shepherds, the balance was redressed elsewhere, where farmers saw their fields uprooted to allow the large-scale rearing of sheep for the looms of Flanders and the fairs of Champagne. The monasteries were among the first to enclose arable land for commercial sheep farming.

These developments did not happen overnight; most were small, half-perceived intrusions and gradual alterations to a pattern of life that must have seemed immutable. The mass of people inhabited a world where wolves still howled in the wil-

Sleeping men beneath the boughs of a yew in a fifteenth-century woodcut *(far left)* represent the funereal associations of the tree. Commonly found in graveyards, the yew was believed to bring dire misfortune or premature death to any who slept or ate in its shade. In contrast, the hawthorn that bedecks the boat in the sixteenth-century Flemish watercolor on the left symbolized hope and fertility. Its blossom-laden branches were carried into houses on the first day of May as protection against maleficent spirits.

derness; where tales of heroes battling dragons, giants, and demon-enchantresses seemed as plausible as village gossip; and where life spans were shortened by epidemics, malnutrition, dirt-borne infections, and the perils of childbirth. Human life and the forces of nature remained intimately intertwined. Yet those who lived close to the land were not the passive victims of their environment: They possessed practical knowledge—of the medicinal uses of plants, the behavior of animals, and the signs of changing weather—that helped them survive and prosper. They knew how to read the world around them as fluently as scholars could scan a text.

Time was reckoned loosely: There was little need to compute a unit as small as a minute. The instruments available to mark time—hourglasses, water clocks, or sundials—were cumbersome or unreliable. Most of humanity tracked the passage of days by the disappearance or return of light; of hours by the shifting position of the sun and the stars or, for clerics, by the sequence of daily prayers; of weeks by the summons to Sunday Mass; and of seasons by the liturgical calendar, the changing weather, and the cycle of agricultural tasks. The landscape itself was both clock and calendar. "We were sitting under the elm," testified one fourteenth-century French peasant during a court case, "at the season when elms have put forth their leaves."

Space was even more nebulous a concept than time, and for most people, the world was delineated by the distance that could be covered in a day. Despite improved roads, travel for those whose lives and livelihoods demanded greater movement—peddlers, sailors, merchants, pilgrims, wandering monks, mercenaries, court officials— was slow and arduous. A well-laden oxcart might cover a little more than three miles in an hour. A convoy of merchants or a noble with his retinue would be pleased to cover about twenty miles between morning and evening; an unencumbered messenger riding a fast horse could, with luck, double that distance.

Despite these limitations, horizons were widening. By the twelfth century, the scholars, merchants, and clerics of Christian Europe had come in contact with a society

A detail from Pieter Brueghel's *Peasant Wedding* (left) shows the last sheaves of the harvest hanging on a tavern wall. Kept through the winter, they were believed to ensure a good crop the next year. In a carving from a fifteenth-century French misericord (right), a blacksmith shapes a horseshoe, whose supposed magical powers derived from both its crescent-moon shape and its material: The magnetic properties of iron and its arcane smelting and forging processes caused this metal and those who worked it to be regarded with awe.

that was in many ways far more sophisticated than their own. The encounters were not always peaceable: For many European Christians, their first perception of the Islamic world came during the era of the Crusades, a century of religiously inspired military campaigns intended to wrest the Holy Land from Turkish and Arab control.

The Muslim communities of the eastern Mediterranean region, North Africa, and Spain had suffered far less from the political chaos and cultural disruption that had afflicted Christian Europe after the collapse of Rome. Since the dramatic spread of the Islamic faith in the eighth century, the Arabic-speaking community had provided a hospitable climate for the preservation and recovery of classical science and philosophy. In the Christian West, the scientific and philosophical works of Aristotle, the geography of Ptolemy and Strabo, and the mathematical studies of Pythagoras had survived mainly as fragments, or through the distorting glass of commentaries upon long-lost texts. Within the Arab world, prolific translators of Greek originals provided a rich intellectual inheritance, one that was rapidly expanded and advanced.

In their understanding of geography and astronomy, their techniques of map making, their navigational skills, and their firsthand knowledge of remote regions, medieval Arabs seemed to live in a different world from their Western contemporaries. They had assimilated the idea that the earth was spherical, instead of flat, and divisible into lines of latitude and longitude. Their detailed navigational charts were more accurate than any previous attempts, and Arab sailors were the first in the West to learn the use of a compass, possibly borrowing this device from the Chinese.

Armed with these skills, the Arabs built a trade network that extended from the Indian Ocean through the Persian Gulf, up the Mesopotamian waterways to Baghdad, and across the overland caravan routes to the Mediterranean. The wares they carried made their way to European markets, where eyes widened at Chinese silks, Indian spices, tropical hardwoods, and the improbable fruit called the coconut. Less glamorous but more powerful in its impact was another Eastern import that Muslims

A group of morris dancers entertains onlookers in this detail from *View of Richmond Palace*, an early-seventeenth-century painting of the Flemish school. The dance evolved from the joint influences of ancient fertility rites and a Moorish dance known as the morisca. The man dressed as a woman represents Mother Eve, while the hobbyhorse rider is descended from Saint George, the dragon slayer. It was believed that the stomping boots of the dancers and the hoofs of the hobbyhorse had the power to rid the ground of the winter frost.

bestowed on the Western world: In the twelfth century, they mastered the technique of making paper. Far cheaper to produce than parchment, paper turned books into tools instead of treasures and sowed the seeds for an intellectual renaissance.

Western scholars did more than wonder. They set to work studying Arabic, so they might translate this fund of knowledge into Latin, the language common to learned Christians. During the twelfth century, one scholar, Gerard of Cremona, translated an impressive collection of seventy-one works: twenty-one on medicine, seventeen on geometry, twelve on astronomy, eleven on philosophy, four on geomancy, three on alchemy, and three more on logic. In a memoir, one of Gerard's students explained that his teacher, having learned all he could in the Christian academies, traveled to Toledo, in the heart of Arabic Spain, where he studied the Arabic language. "In this way, combining both language and science, he passed on the Arabic literature in the manner of the wise man who, wandering through a green field, links up a crown of flowers, made not just from any, but from the prettiest."

Muslim and Christian minds met not only in the frontier region of Spain but also along the trade routes controlled by Constantinople and Venice as well as in Sicily, where King Roger II ruled a twelfth-century court that mingled the culture, architecture, and learning of Islam and the West. Crusaders, traveling to or from the Holy Land by sea, used Roger's island realm as a staging area. They returned home inspired by the learning of his court, by the Greek domes and Byzantine mosaics of his palaces, and by the beauty of his estates and gardens. One Italian visitor reported that "in order that none of the joys of land or water should be lacking to him," Roger had built a sanctuary for birds and beasts and had stocked its waters "with every kind of fish from diverse regions." Near his capital he had created a walled park that was "shaded with various trees and abounding with deer and goats and wild boar."

Northern princes took note, and copied. Robert of Artois, returning home to the northernmost part of France after a thirteenth-century Crusade, constructed a garden that compensated for the unattainable balmy, perfumed breezes of its Mediterranean prototypes by a dazzling array of fountains and ingenious moving statuary. Other travelers took cuttings of unfamiliar fruits, vegetables, and edible flowers from Arab gardens: spinach, artichokes, oranges, lemons, apricots, and saffron. And Christian travelers also absorbed the underlying philosophy of the Arab horticulturalists: The garden was a version of the oasis, an enclave where nature was tamed and transformed. In the north, the less-orderly world outside the garden's wall might be forest or scrub instead of desert sand dunes, but the principle was the same.

This new regard for the aesthetic beauties of nature testified to a more general change in attitude within Europe. In the early centuries of the Christian Era, the teachings of the Church had directed attention heavenward: The next life, not this one, was the proper focus for concern, and if plants or animals or the land had any significance, it was as symbols of a higher spiritual truth. The pelican, for instance, which allegedly pierced its own breast to feed its young on the blood that gushed forth, was a symbol of Jesus Christ, sacrificing himself for humankind; the crab, with its sideways gait, embodied fraud or deceit. The earth was a textbook, written by God, interpreted by priests. Such beliefs persisted throughout the Middle Ages, but they were joined, and eventually supplanted, by an interest in the natural world for its own sake, a growing enthusiasm for accurate observation and information.

The thirteenth-century Franciscan friar Bartholomaeus Anglicus was one of several encyclopedists who met this need with sweeping compendiums of knowledge that combined the factual and the fanciful in almost equal measure. In a passage on the crocodile, Bartholomaeus repeated the wisdom that was already old when it was retailed by the Roman author Pliny the Elder, noting that "if the crocodile findeth a man by the edge of the water or by the cliff, he slayeth him if he may, and then he weepeth upon him, and swalloweth him at the last." Bartholomaeus's description of Finland dwelled on its inhabitants' propensity for witchcraft, especially in the matter of conjuring winds for visiting sailors, while an inventory of human organs assigned to each its function: "By the spleen we are moved to laugh, by the gall we are angry, by the heart we are wise, by the brain we feel, by the liver we love."

Equal to the intellectual passion for acquiring knowledge of the world was the passion for arranging this information into orderly numerical categories. The composition of all animate and inanimate objects was based on the presence or absence of each of the four elements: earth, fire, air, and water. The appearance, personality, and health of every human being depended on the dominance of one of the four humors: the sanguine, the phlegmatic, the choleric, and the melancholic. Good or bad fortune was intimately related to the influence of the twelve signs of the heavenly zodiac; spiritual salvation required the embrace of the seven cardinal virtues and avoidance of the seven deadly sins. The theologian Albertus Magnus divided the world into seven climates and showed how their peculiarities of weather and landscape affected the development of human beings and animal species. He explained, for instance, that the northern-dwelling races, such as the Goths, the Danes, and the Slavs, had pale skins because of the cold climate they lived in: The frigid air constricted their blood flow and turned their bodies white.

An age that was dominated by feudal and clerical hierarchies imposed the same imagery of rank and order on the universe. Heaven and earth were joined in a great chain of being, with God at the top. Human beings, according to their social status, took their places in the middle; then came animals, plants, and inanimate objects. Just as men and women existed to serve God, and peasants existed to serve their lords, so lower forms of animal, vegetable, and mineral life existed to serve humankind.

But there were chinks in this system, as exemplified by the influential teachings of Francis of Assisi, who died in 1226. Saint Francis, founder of the order of wandering friars who came to be known as Franciscans, insisted that all living creatures were worthy of kindness and respect. He attracted mockery by his practice of preaching to congregations of birds, on the grounds that they too deserved to hear the word of God, but he ultimately came to be widely admired. He composed canticles of praise to Brother Sun, Sister Moon, Brothers Wind and Fire, and Sister-Mother Earth, "who sustains and governs us and produces fruit with colorful flowers and leaves!"

The dismissal of earthly pleasures as spiritual traps was yielding to a more sensual, if still reverent, celebration of the natural world. Where the fiercely ascetic old desert fathers might have sniffed brimstone on every summer breeze, or seen the blooming rosebush only as a snare of Satan, even a pope could now enjoy a ramble along the ancient Appian Way out of Rome to Lake Nemi, remarking that "Nature, who is superior to any art, has made that road most delightful." On April 26, 1336, the Italian poet Petrarch with his brother and two servants climbed Mont Ventoux, about thirty miles northeast of Avignon—the first recorded ascent of a mountain purely for the enjoyment of it. "I could clearly see the Cévennes to the right," Petrarch wrote to a friend on the same day, "and to the left the sea beyond Marseilles and Aigues-Mortes, all several days' journey distant. The Rhone itself lay under our eyes."

A Flemish miniature from a late-fifteenth-century edition of *The Romance of the Rose,* a narrative poem written two centuries earlier, shows many distinctive features of the medieval garden, such as elaborate latticework, raised flower beds, running water, and a variety of fruit trees, some of which were introduced to Europe after the Crusades. The rose, which in the poem symbolized the fulfillment of love, also originated in the East. Mirroring the shape of a monastic cloister, the enclosed garden offered not only privacy and shelter from the workaday world but also an idealized setting in which the sensual pursuits of music and love acquired a religious significance.

The lavish decoration of the new churches and cathedrals reflected this shift in attitude. Saint Thomas Aquinas, prime philosopher and theologian of the age, declared that "works of art are only successful to the extent that they achieve a likeness of nature." Sculptors and woodcarvers concentrated on the creation of faithful copies of their favorite flora: snapdragons, ivy, cress and celandine, oak leaves, strawberries, ferns and broom, stone replicas of vine shoots twining around the doors. Fantastic creatures—the dragon and the basilisk, for example, with its body a splice of bird and serpent—were juxtaposed with those that had been faithfully copied from living models. Villard de Honnecourt, a celebrated thirteenth-century French architect, kept sketchbooks full of his meticulous drawings: swan, cat, bear, lobster, dragonfly, and a chained lion from a noble's private menagerie, its portrait proudly labeled: "Take notice that this was drawn from the life."

The Gothic architecture that reached its peak in the churches of thirteenth-century France, Flanders, and England carried within it memories of the great forests that had now been tamed. Long naves and pointed arches recalled the vistas seen through avenues of trees and the shafts of light piercing tangled branches. The pagan spirits of the northern forests—the guardians of trees and streams, the Green Man of legend—penetrated the Christian sanctuaries in the form of smiling grotesques, leaves sprouting from their beards and hair, that peered from stone medallions above church porches or emerged from the carved, polished wood of choir stalls.

The building of the great Gothic cathedrals marked the high point of an era of optimism, intellectual energy, and an expansion in all manner of human enterprise. But by the beginning of the fourteenth century, the population increase that had fueled these developments was itself becoming a source of strain. Many parts of Europe stood virtually at the saturation point; much of the countryside was overpopulated, with too many mouths for the available ground to feed. Soils in some areas were depleted, prices of foodstuffs high, and the wages for labor low.

As early as 1258, parts of England experienced the first in a series of food shortages, and by the first decade of the new century, Germany was hit by famines nearly as grave as those told of in grim detail by Radulfus Glaber. And to add to the suffering, the climate that had been so temperate since the millennium began to worsen. Longer and harsher winters caused crops to fail; remote northern or mountain settlements suffered from cut-off communications, languished, and sometimes died out.

The gravest crisis of the age began in the 1320s, when the plague-carrying bacterium, *Pasteurella pestis,* emerged from ratholes somewhere in the Gobi Desert in Mongolia, and—carried by the rat flea—entered China with its host. From China, the infection spread along the East-West trade routes, conveyed by the rats inhabiting the caravansaries and delivered to towns and markets with the bundled merchandise. By 1346, the plague rats had reached the Crimea, on the Black Sea, whence they traveled by ship to European ports and then inland along roads and waterways. Ironically, the spread of the epidemic was facilitated by Europe's economic and technological advances: Improvements in shipbuilding, resulting in vessels that were seaworthy year-round, brought both enhanced trade and the Black Death.

In densely populated communities, the germs were spread not only by infected fleas and rats but by the coughs and sneezes of people already infected. Even the most isolated settlements were vulnerable: A single plague-carrying rat on a river barge or a lone infected traveler could bring the pestilence, with its erupting sores, rapid-fire fever, and respiratory distress. Not all parts of the Continent were equally afflicted: The sparsely populated grain belt of Bohemia and Poland suffered less than the highly urbanized regions of northern Italy and Mediterranean France. But the overall toll was horrendous: At least one-third of Europe's population perished, either in the first onslaught at midcentury or in the recurrences of the 1360s and 1370s.

It would be 200 years before the population of Europe returned to preplague levels. But despite the devastation and the social tensions, Europe did not slide into stagnation. Economic and social recovery was slow, but steady. With fewer mouths to feed and fewer laborers to assist in the feeding, farmers cultivated only the best and most fertile tracts of land, which would provide the highest yields with the lowest expenditure of effort. Marginal land was allowed to slip back into wilderness. A shortage of workers caused earnings to rise, and in many districts, the diminished population found itself enjoying a much higher standard of living than ever before.

Moreover, those who survived or were born after the Black Death enjoyed freedoms and opportunities unknown to their forebears. The old feudal structures were crumbling: Peasants were no longer serfs, owing labor and obedience to their lords, but tenants, paying rent and free to move on if they saw better opportunities else-

THE NOBLES' PURSUIT

On Christmas Day 1251, Henry III of England presided over a banquet that included 630 deer, 200 boars, and 2,100 partridges. This excess of meat was both a ritual expression of the king's wealth and symbolic evidence of his prowess as a hunter—for in medieval Europe as in most preindustrial societies throughout the world, a nobleman's quality was judged above all by his courage and skills in the chase.

Hunting served many practical purposes: It provided training for warfare, rid settlements of predators, and contributed—through the creation of royal parks—to the conservation of woodland. But for many it was an end in itself, a specialized art form whose practice required an intimate knowledge of the behavior of animals. This was especially the case when the hunter employed trained animals to supplement his own weapons—a tradition that, as the illustrations here and overleaf indicate, had roots in many cultures. The Indian miniature on the right, painted around 1600, shows Emperor Akbar hunting deer with a cheetah; at the lower left, a blindfolded cheetah waits on a cart to be released.

In a tomb painting dating from 1400 BC, an Egyptian, who is accompanied on his outing by his wife and daughter, hunts marsh birds from a papyrus raft. The hunter holds three decoy herons in his right hand, while his retriever cat has cleverly trapped three birds—one in its mouth, one in its front claws, and a third pinned down by its tail.

A nineteenth-century Japanese ivory statuette depicts a falconer launching his bird toward its prey. Widely practiced in Asia from the eighth century BC, the sport of falconry was introduced to medieval Europe by Crusaders and merchants who were returning from their journeys to the Orient.

A hunter pursues a deer *(above)* with the help of hawk and hounds in a twelfth-century ironwork decoration from Sweden, where hawks were trained to peck out the eyes of their prey. On the left, a sixteenth-century tapestry depicts hounds savaging a boar while their master, the Holy Roman Emperor Maximilian I, prepares to deliver the fatal thrust with a sword specifically designed for the sport.

In a tenth-century Chinese drawing, hunters returning from the Asian steppes share the saddle with their Saluki hounds. The Saluki, a breed domesticated as early as 7000 BC, was used for hunting in ancient Mesopotamia and Egypt.

where. Towns, though no longer bursting at the seams after the depredations of the plague, were becoming the true centers of economic life and power, not only in the Mediterranean lands but also in the north. The noble lord of the rural manor now had a potent rival in the person of the urban merchant, who preferred trade missions to Crusades, felt more at home in the counting house than on the battlefield, and showed his loyalty to his monarch by making loans instead of feudal vows.

Nor was the advance of science and technology slowed by the years of plague. By the early 1400s, new Latin translations of important classical authors—such as Ptolemy on geography, Euclid on geometry, and Vitruvius on architecture—made these writers and their ideas more widely accessible and, therefore, far more influential. Gunpowder, acquired from the Chinese through the agency of the Arab world, transformed the logistics and the machinery of warfare. The mining industry, responding to the increased use of metal coinage and the growing demand for iron tools, became more productive through a battery of new devices: suction pumps, cogwheels, and systems of gears, powered by horses or by water. Even the measurement of time had progressed: The first weight-driven clock had been installed in Milan in 1335, and by the beginning of the fifteenth century, it was a poor city indeed that could not boast a similar dial on the tower of its cathedral or town hall.

If the world had changed, so too had the ways of looking at it. Nowhere was this

clearer than in the transformation of the visual arts. Northern painters of the Flemish school scrupulously reproduced the hunting parks of local gentry and the features of both known and imaginary landscapes as the background to religious and secular scenes. In the brilliantly detailed illustrations to the book of hours created for Jean, the duke of Berry, the artists depicted the turning year through the actions of men and women within a landscape that bore all the marks of human intervention and control. And in Italy, painters experimented with proportion to reproduce the human form in a way that would be not only mathematically accurate but aesthetically pleasing.

Subject matter developed accordingly. The religious scenes that still dominated artistic production were no longer flat, impersonal images but dramas played out by human beings who wept, bled, suffered, or rejoiced as they might have done in life. To further these efforts to imitate reality, artists such as Brunelleschi developed the use of linear perspective: the technique for creating a picture in which the size, distance, and relative positions of all objects faithfully imitate the spatial relationships seen in a slice of the real world, as viewed by someone looking in through a window.

While painters studied methods for making line, light, and shadow into plausible copies of the subjects they represented, a complementary breakthrough was taking place during the 1440s in a workshop in the German city of Mainz, where Johannes Gutenberg and his colleagues developed a method of printing books by means of movable metal type. The art of printing was not a new discovery: The Chinese had been reproducing texts on a form of mechanical press since the ninth century, but the technique had apparently never traveled farther west than Turkistan in central Asia. Once Gutenberg introduced movable type to the West, however, it spread more swiftly, and more beneficently, than the plague. By 1474, books were coming off presses in Paris, Venice, Lyons, Naples, and the Polish city of Kraków. By 1500, print shops in 236 European towns and cities had turned out an estimated 20 million volumes—at a time when the population of the Continent was only 70 million.

Linear perspective and movable type together promoted a revolution in the transmission of information. Ideas were now accessible to those who could master the skill of deciphering letters; illustrations, maps, and diagrams could be easily and clearly reproduced. Intellectual stimulation was no longer the preserve of clerics in their monasteries and patrons in their palaces, poring over rare, hand-copied texts. Facts, discoveries, and contentious theories—on medicine, mathematics, geography, chemistry, botany, and astronomy—flew back and forth among Europe's scholars.

Some of the new texts posed dangerous challenges, and the guardians of religious orthodoxy would do their best to ensure that the books, and their authors, were consigned to the flames. The fears of such conservatives were understandable: Christians could now read for themselves the words of the Bible and make their own interpretations of religious truth. But the flow of ideas could not be stopped. Those who could read—and their numbers were growing at a prodigious rate—had no reason to remain a passive recipient of someone else's view of the world. Everything in creation was open to question, and every question was soon to be asked. The printed word, that most arbitrary set of abstract symbols, had opened the floodgates. The book of nature was about to be rewritten.

A sign from a Venetian workshop shows men at work on vessels in the Arsenal, the greatest shipbuilding complex of medieval Europe, which took its name from an Arabic word meaning "house of industry." The Arsenal supplied Venice with galleys for warfare and a merchant fleet of more than 3,000 ships; the latter included light inshore caravels and square-rigged vessels sturdy enough to sail into international waters. The growth of sea trade within Europe, dominated by Venice and Genoa, stimulated improvements in ship design that made possible the voyages of discovery to the New World.

EXPANSION AND EXCHANGE

5

In March 1519, runners brought to the Aztec emperor of Mexico news of outlandish invaders who had arrived on the shores of his domain. "The strangers' bodies are completely covered," reported the astonished messengers, "so that only their faces can be seen. Their skin is white, as if it were made of lime." And even more strange: "Their deer carry them on their backs wherever they go. Those deer, our lord, are as tall as the roof of a house."

For Emperor Montezuma II, only two explanations were possible: Either his messengers were liars, or his territory was being invaded by supernatural beings. In fact, neither of these suppositions was correct—but the Aztecs had never seen white-skinned men before, nor had they seen horses. And the paralyzing fear induced by the appearance of these wondrous new species so sapped the Aztecs' will to resist that within two years their massed armies had been vanquished by a force of just a few hundred Spanish conquistadors.

The Spaniards had other advantages besides their appearance. They possessed swords of cold steel and cannon that could inflict damage on a scale immeasurably greater than that of the Aztecs' own spears and arrows. "A thing like a ball of stone comes out of its entrails," reported the Aztec envoys of this terrible weapon. "It comes out shooting sparks and raining fire." During the following centuries, however, the chief means by which European invaders dominated both North and South America was not military hardware but a biological arsenal of diseases, animals, and plants. Isolated from the continents of Europe and Asia for thousands of years, the peoples of the Americas lacked immunity to the invaders' ailments. And in the wake of the Spaniards' horses came a menagerie of other animals wholly new to America: cattle, pigs, sheep, and goats among others, whose voracious appetites reduced many indigenous plants to the point of extinction. Finally, like a vanguard of skirmishers, Europe's weeds preceded the inexorable advance of Europe's wheat, barley, and other crops, which in turn fueled a rapid increase in the numbers of new animals and people alike. By 1800, much of what had at first been to the Europeans an alien wilderness had been transformed into a home away from home.

The pace of this ecological invasion, although it came to possess a momentum of its own, was forced by very human factors, not the least of which was the Europeans' assumption that the riches of the natural world were theirs to exploit and that success could be measured in terms of economic profit. And sea power gave Europe the opportunity to impose a similar form of biological colonization wherever they made landfall. In some regions, the graft did not take; but wherever climatic conditions permitted European flora and fauna to flourish and the land was not densely populated by resistant peoples, the story of the Americas was successfully replayed. Some indigenous plants and animals traveled in the opposite direction, it is true, but these,

A detail from a canvas painted by the Dutch artist Jan Mostaert in the 1540s shows American Indians on a craggy outcrop hurling down stones on an advancing column of Spanish conquistadors. The painting may have been based on an eyewitness account of a battle fought in 1540 in present-day Arizona between Spanish soldiers—who were tracking a rumor of gold—and Zuni Indians. Although Mostaert captured the true savagery of the conflict between New World inhabitants and Old World invaders, his vision of a pastoral landscape where sheep and cattle calmly graze in a well-wooded field accorded more with European expectations and artistic conventions than with reality.

too, contributed to the long-term advantage of Europe. They supplemented Europe's diet, and they fueled intellectual and scientific advances that allowed the colonial powers to extend their reach still farther.

The advantages that the Europeans were able to exploit so successfully were the product of thousands of years of evolutionary forces. Approximately 200 million years ago, the slow grinding of the great rock plates that make up the outer crust of the earth began to tear apart what was then the planet's single landmass, dividing it into separate continents on which plants and animals evolved in dissimilar ways.

Captured in the grid of a map from the first atlas of the whole world, published in Antwerp in 1570, the recently discovered continents of North and South America are suspended with Europe, Asia, and Africa above the hypothetical landmass of Australia. Following the transatlantic voyages of Christopher Columbus in the 1490s, peoples on both sides of the ocean were forced to revise their understanding of the world. To the Indians of southwest America who painted on a rockface the Spanish cavalcade shown above, even the horses that bore the invaders were utterly new.

Humans, too, developed their technological abilities at different paces. As a consequence, when the Europeans came face to face with their long-lost cousins, the parties scarcely recognized their kinship. The world was suddenly larger and more diverse than anyone had recognized, and it challenged every convention by which it had been understood.

Some historians believe that the first humans to reach the Americas probably arrived before 13,000 BC, during a period when the continent was linked to Siberia by a land bridge laid bare by falling sea levels during the last Ice Age. South of the slowly shrinking ice, they found an enormous new territory, free from human rivals and abundantly provided with the big game they were so efficient at killing with their spears and flint arrowheads. These peoples' numbers increased as they spread southward; around 10,000 BC, they had established themselves everywhere from the edge of the Arctic to the distant tip of South America. But by that time, the melting glaciers of the dying Ice Age had raised the sea levels and closed off the bridge from Siberia. Like their fellow pioneers in Australia, who were also isolated by rising sea levels, the new Americans were left to develop in their own manner.

They made the most of their resources. The American continent had been a separate biosphere for millions of years, and the newcomers were largely untroubled by the irksome parasites and predators that had evolved alongside their primate ancestors in the Old World. In similar fashion, the animals of this continent had never evolved a defense against primate rapacity, and they never would, for within a short time most of the larger species—including mammoths and the original American horses—had become extinct, at least in part because of large-scale hunting. But the people of the Americas adapted their weapons for smaller game and thrived on what they could gather from a fertile land. Around 5,000 years ago, those living in the most crowded part of the continent—approximately modern Mexico—began to cultivate wild corn. This crop evolved over the generations into America's most important native food plant, supplemented by squash and beans, which were both richer in protein. By 1500 BC, village-size settlements were common; by AD 500, Mexico supported some notable cities and the complex, organized society that went with them. Although the settlers failed to put the wheel to productive use and had no knowledge of iron, they were learning the skills of metalworking, especially in gold and silver. The human invasion of the Americas could be reckoned a success.

The far more numerous humans left behind in the Old World, however, had done even better. To begin with, they adopted agriculture thousands of years before their American relatives, and whereas the Mexican protofarmers took many centuries to increase corn yields to a level that might support a civilization, the wild wheat the Eurasian cultivators began with was already heavy with plump and nutritious grain.

At least as important as the agricultural advancements, the domestication of key animal species was a late-developing skill. It is believed that the first humans who entered the Americas were accompanied by the dogs that they had successfully tamed. In later millennia, Old World humans domesticated a roll call of species that probably included cattle, sheep, goats, horses, pigs, buffalo, reindeer, and even honeybees. But when the Americans, too, began to learn the technique, they found few suitable subjects. The scarcity may have been partly a consequence of their early success as hunters. At any rate, against the formidable array of species that served Old World humanity, the Americans could muster only the llama, the guinea pig, and a few varieties of poultry.

Their animals gave the people of the Old World enormous advantages, of which a ready supply of meat, hides, and fertilizing manure were only the most obvious. Plows that were pulled by horses and oxen could cultivate the land well beyond human strength to dig, and pastoralists could live off their herds in marginal terrain that would support cattle but not crops. Some Old World humans took their adaptation to domestic animals a step further: The Europeans in particular had begun to thrive on a diet of milk right through adulthood; few Asians and no native Americans could tolerate milk after infancy.

The Old Worlders also learned to live with the disadvantages that came from the propinquity of human and beast—and here they would find their most powerful weapon against their distant cousins. Most of their domestic animals carried microorganisms and parasites such as fleas that could cross the species barrier and afflict humans, usually with far greater severity than they did their original hosts. But given sufficient time, the survivors evolved at least partial immunity to many of these infections. Besides, as Old World cities developed, with many centuries' head start over the nascent civilizations of America, the most dangerous infections—smallpox, for example—became endemic in their crowded populations. Disease still killed many thousands of people, but usually in childhood; adults were left with an armory of antibodies. The people of the New World, who had never had to deal with anything like the Old World's pool of lethal pathogens, had no such defenses. Compared with the disease-hardened men and women of the earth's central landmass, they were biological innocents.

For several thousand years, the Americas remained isolated by the quarantine of distance. But in the fifteenth century, the peoples of Europe, led by the Spanish and

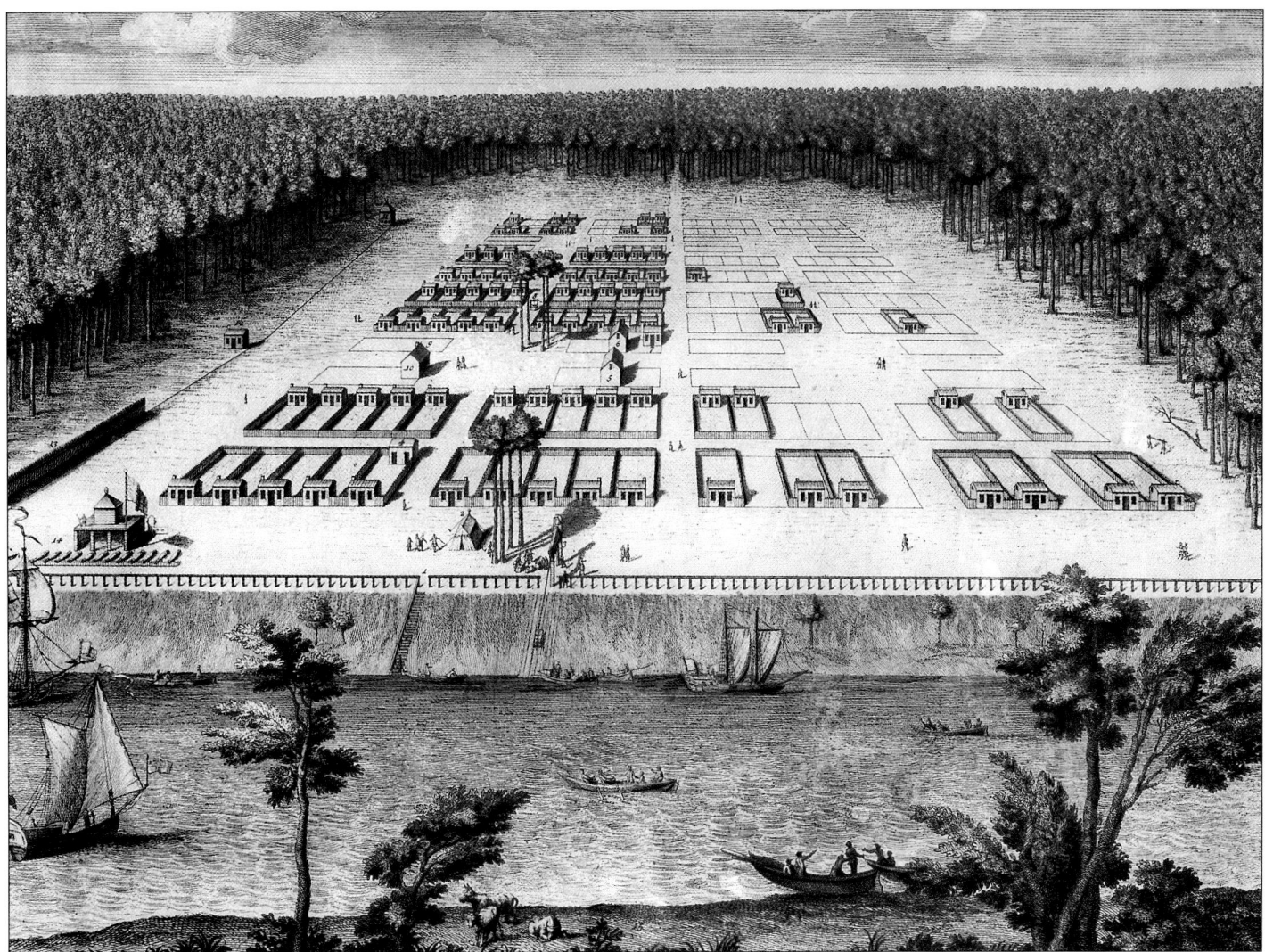

A vision of order imposed on an alien wilderness, this engraving based on a drawing by Peter Gordon shows the uniform building plots of the settlement of Savannah in the colony of Georgia as they stood at the end of March 1734. The town had been founded in the previous year by a British member of Parliament, James Oglethorpe, whose tent is pitched between the riverbank and the first houses. The colonists, whose landholdings were increased from 500 to 2,000 acres in 1740, were encouraged to produce wines, spices, and silk to relieve Britain's dependency on more expensive sources. In 1819, Savannah was the embarkation point for the first steamship to cross the Atlantic.

the Portuguese, steadily gained a knowledge of the pattern of the transoceanic winds, without which no one could hope to cross an ocean and return alive. And they were equipped not only with ships capable of carrying them around the world but also with the feelings of cultural superiority that would-be conquerors needed just as much as their weapons and charts.

By the 1440s, the Portuguese were settling the uninhabited islands of the Azores and Madeira, transporting their own plants, animals, and slaves there and raising sugarcane for Europe's insatiably sweet tooth. Farther afield, their ventures to the Indian Ocean brought profits that more than compensated Portuguese mariners for their hard apprenticeship learning the ways of the ocean winds. Portugal's Iberian Peninsula neighbor, Spain, became engaged in a one-sided genocidal war against the Stone-Age Guanche people who lived on the Canary Islands. And when the Genoa-born mariner Christopher Columbus began begging and wheedling his way around the courts of Europe to finance a voyage westward from the Canaries, it was the riches to be gained from a new route to China that eventually persuaded Ferdinand and Isabella of Spain to back his venture.

With his ninety men and his three small ships, Columbus made an unspectacular landfall in October 1492 after an unexciting voyage that lasted a mere thirty-three days, an almost absurdly easy cruise compared to the wide, pelagic swoops of Portugal's mariners who were bound for India. Columbus believed up to the time of his death in 1506 that he had found his longed-for route to China. But, of course, he was not in the East Indies or anywhere near them—his ships had touched on the Bahamas, and he had discovered not a backdoor route to the other end of Eurasia but an entirely new continent.

The Spaniards did not delay in exploiting the breach that Columbus's voyage had opened. By the beginning of the sixteenth century, a horde of adventurers had followed in Columbus's wake, and the Iberian appropriation of the island of Hispaniola was almost complete. For the first few years, the Spaniards contented themselves with little more than pillage, seizing gold and slaves to take home with them. Soon, they settled into serious exploitation, growing sugar in a successful attempt to match the lucrative production of Madeira and the Azores. The pigs, cattle, and even horses Columbus had brought were breeding with enthusiasm, rooting out and trampling down what remained of the original ecology. Many of the islanders were killed; most of the rest became slaves on the sugar plantations their new masters created. To eke out the labor force, the first African slaves were hauled across the Atlantic in 1505. And in 1518, another immigrant arrived: the smallpox virus. According to one witness, barely 1,000 of the original Amerindian population of Hispaniola survived. A scant generation after Columbus's lookouts first saw the island loom above the Atlantic swell, its original people, animals, and plants had been virtually annihilated—and rapidly replaced.

THE NATURALISTS' RECORD

Diamond python, painted in Australia around 1790

The contents of the pouch held by the Swedish botanist Carolus Linnaeus in the portrait on the right included pens, brushes, and a microscope. No less important than guns and plows in the European appropriation of the Americas and Australia, these tools were the instruments by which newly found flora and fauna were observed, recorded, and caught in the expanding web of Western science.

The classification system for plants devised by Linnaeus enabled botanists to place new discoveries in named categories and so establish their relationship to other species. Other naturalists, too—including those shown on the following pages—contributed to a rational, objective understanding of the natural world by their interpretation of the data they collected.

On the offshore islands, smallpox merely finished a job the Spaniards already had well in hand. On the mainland, where settlement had begun in 1509, it was decisive. One year after the tiny band of conquistadors led by Hernán Cortés launched their against-the-odds invasion of the huge Aztec empire, the sickness crossed over along with reinforcements from Hispaniola. The extra humans were not nearly enough to turn the tide, but the virus one of them carried more than made up for their lack of numbers. Within a few months, a raging smallpox epidemic, to which Cortés and his men were almost all immune, had killed perhaps half the native population. The death rate was highest among young adults—a grim commonplace in an encounter between an infection and a human group with no established immunity against it—and the survivors were utterly demoralized. Victims of invaders who appeared to have divine or diabolic power on their side, the Mexicans soon found themselves serfs of the viceroyalty of New Spain.

The disease spread southward, ravaging the Inca dominion of Peru and opening the way to another improbable victory a decade later, when Francisco Pizarro added the viceroyalty of Peru to the Spanish crown. The affliction ran still farther, burning like a forest fire in a drought, the spark of new outbreaks often borne by refugees who carried in their own bodies the death that they imagined they had left behind. And following the pestilence came the empire of Spain, relentlessly expanding down the length of South America, with a plethora of other lethal illnesses—including measles, mumps, and diphtheria—seemingly waiting for those who smallpox failed to exterminate.

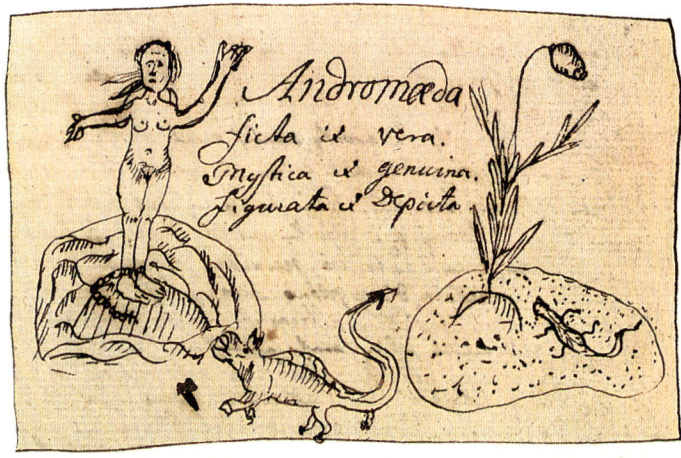

In a 1737 painting *(left),* **Linnaeus is dressed in the traditional garb of Lapland, where he traveled to study plants in 1732. A sketch from his notebook** *(above)* **shows the Greek mythological figure Andromeda and the flowering marsh plant that he named after her.**

Lobster krill, painted off the coast of Patagonia in South America in 1769

Infectious disease was the single most powerful weapon the Europeans had against indigenous peoples. Against indigenous ecologies, their plants and livestock were just as effective, usually without their owners' active intervention. The most startling example was in the vast plain of the pampas, the grasslands that extend across the area of modern Uruguay, northern Argentina, and southern Brazil. The Spanish abandoned a failed settlement in the region early in the sixteenth century, but although the colonists went home, their animals did not. They ran wild and prospered, for they had no competition: There had been no grass-eating ruminants in the region for thousands of years. When the city of Buenos Aires was successfully founded at the second attempt in 1580, there were already substantial herds of feral cattle, and from then onward, bovine population growth was explosive. The local grasses, not adapted to withstand steady grazing, were soon wiped out. But cattle numbers—there were plenty of wild horses, too—continued to rise because the grasses were quickly replaced by their European analogues, which had evolved over millions of years to thrive despite heavy cropping. In effect, the partnership of plant and animal Europeanized an immense tract of South America with no more human intervention than was necessary to transport them across the ocean.

The colonization of North America had to wait until its English, French, and Dutch protagonists had developed enough sea power to crack the Iberian near-monopoly on transatlantic communications. After a few of their undersupplied settlements in the sixteenth century had succumbed to malnutrition and Indian attack, the efforts of these countries to colonize the New World began in earnest in the 1600s, by which time smallpox and other diseases had already prepared the way. Spanish forays up the Mississippi River had exposed the indigenous population to deadly infection, and the new arrivals brought diseases of their own. A French outpost in modern Nova

A 1795 cartoon satirizes the rise to social prominence of the British naturalist Sir Joseph Banks by showing him as a butterfly evolved from a grub. Three pictures *(opposite)* illustrate the stages by which some of Banks's discoveries passed into common knowledge: the specimen plant collected in Botany Bay, Australia, and named after Banks; an on-the-spot sketch of the same plant executed on board ship in 1770; and a finished watercolor prepared in England for publication as an engraving.

Spiny anteater, painted by a convict artist in Australia around 1790

Scotia noted a disastrous epidemic in 1617, which scorched its way along the eastern seaboard—probably smallpox, but at any rate a preparatory bombardment of microbiological artillery that greatly assisted the landing of the English Pilgrims in Massachusetts in 1620. Farther south, the Indians were less depleted and came close to overrunning the English settlements in Virginia in 1622. But there, too, European pathogens were providing their terrible assistance.

Fighting between English and French colonists held back development at least as much as the rigors of their new environment, but no such squabbles restrained the advance of their allied species. East of the Mississippi, there were no indigenous herbivores, and just as on the pampas, native grasses had not evolved to tolerate constant grazing. Once more, Europe's livestock munched and trod a way clear for Europe's plants, which in turn supported the livestock and the people. White clover—once European honeybees had been introduced to pollinate it—and the so-called Kentucky bluegrass, among the most successful of the imports, did not so much accompany the westward-moving frontier: They were themselves the frontier, crossing the Appalachian Mountains in the eighteenth century ahead of human pioneers. European weeds, the archopportunists of the plant world, thrived on the upheaval. Plantain, for example, earned itself the Indian nickname "Englishman's foot" by appearing everywhere the interlopers trod, and sometimes where they would not tread for a few years yet. For, like the bluegrass and indeed the honeybee—or "English flies" to the Indians—Old World weeds often swarmed ahead of the Old World people who inadvertently introduced them.

The colonists did not depend entirely on their own plants: Where the Amerindians had a useful food crop, Europeans quickly learned to adopt it. If it would grow in Europe, ships carried it back to the homeland. Thus corn—then known as Indian

corn—was soon a staple for intruders as well as natives and quickly became well established in Europe. Along with the tropical root called cassava, it was introduced into Africa, where it probably helped the population to increase despite the depredations of European slave raiders.

The cod fisheries on the Grand Banks off Newfoundland, first reported by the English-sponsored navigator John Cabot in 1498, also proved a boon to the Europeans, and salt cod soon became a staple in parts of the Mediterranean region. Turkeys, too, made the eastward crossing, as did the potato, the most important American food plant of all.

The exchange of plants and animals was mostly a one-way traffic, however. Very few American weeds established themselves in Europe—although hundreds of their European counterparts hitchhiked their way across the ocean to the New World. It was an imbalance that puzzled contemporary botanists, but the reason was not hard to find: Weeds thrive on upheaval, and it was in America, not Europe, that an ecology was being overturned.

On the microbiological level, the Americans may have passed syphilis to the people of Europe: Medical records are ambiguous, but the disease first appeared in virulent form shortly after the first voyages to the New World. But although it claimed many victims, syphilis did not have the population-destroying power of smallpox and the other European contributions to American ill-being; and venereal disease, whether of American origin or not, was to prove one of the final scourges of the Amerindians when their cultures started to disintegrate. Like weeds, the germs were spread on the wings of chaos.

Moorish idol *(right)*, painted off the coral reefs of Tahiti in 1769

Not all the European imports worked directly toward the downfall of the native populations. Horses, for example, which were breeding wild in Mexico, migrated northward in enormous herds into the Great Plains, where the small population of Indian buffalo hunters who lived there took to horsemanship with enthusiasm. The newly arrived animals transformed their lives. Buffalo hunting on foot had provided a marginal existence at best; mounted, the Indians prospered mightily. At the beginning of the nineteenth century, there were at least 100,000 of them, following buffalo herds whose own numbers were probably between 30 and 40 million. By the same time, however, the European colonies that had been established on the eastern seaboard had become a powerful, independent nation of almost five million people. Already they claimed the Mississippi River as their western boundary; and their people, eyeing the vast territories beyond, were increasingly well equipped to seize the new land for themselves.

The Europeans had not reached a position of such advantage without a struggle, but they had been aided beyond their own understanding by their nonhuman allies. It was as if the conquest of America, unplanned and opportunistic though it was, had been controlled by a brilliant general staff with a whole range of species obeying its orders. First, a wave of microorganisms devastated the native people; then a second wave of larger life forms—pigs, cattle, grain—wrought no less dramatic changes upon the environment. It had taken the colonists' ancestors thousands of years to impose their familiar, human-centered ecology upon the Old World; it took only a few hundred to impose it on America.

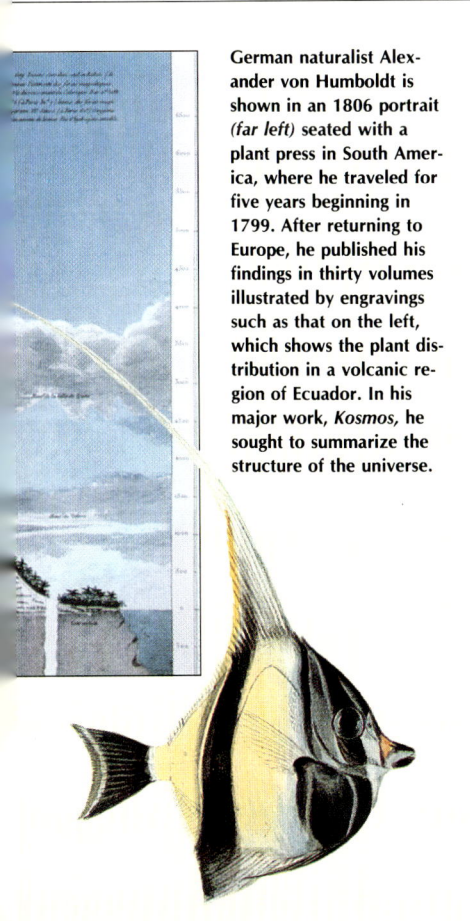

German naturalist Alexander von Humboldt is shown in an 1806 portrait *(far left)* seated with a plant press in South America, where he traveled for five years beginning in 1799. After returning to Europe, he published his findings in thirty volumes illustrated by engravings such as that on the left, which shows the plant distribution in a volcanic region of Ecuador. In his major work, *Kosmos*, he sought to summarize the structure of the universe.

For John Winthrop, the first governor of the Massachusetts Bay Colony in the seventeenth century, the fact that the local population was "near all dead of smallpox" appeared to demonstrate that the "Lord hath cleared our title to what we possess." Governor Winthrop was not the only European who saw the hand of God at work in the American takeover, and indeed many were convinced that divine support would afford the Europeans similar dominion over the rest of the world. But there were limits to the favor that the Almighty appeared to have granted Europe. The colonial accomplishment in America had been resoundingly successful, but building successful colonies elsewhere than in a susceptible ecology whose climate was broadly similar to Europe's own proved a much tougher proposition.

Even in the Americas, the Europeans did not have everything their own way. The profits of the French fur traders, for example, depended not on imported European techniques but on the traders' use of indigenous resources such as birch-bark canoes, moccasins, and snowshoes. In the Spanish domains of Arizona and New Mexico, significant numbers of local people not only survived the European invasion but retained their traditional ways of life largely intact. And although European plants and domestic animals were highly adaptable, wheat could not grow in rain-forest conditions, nor would cattle thrive in a desert.

The colonists were not interested in establishing settlements in the deserts, of course, but the tropics were another matter: Columbus had been greatly impressed by the verdant coast of Venezuela, and explorers of other nationalities echoed his enthusiasm. Attracted by the lush greenery and the apparent prospect of an idyllic lifestyle, many Europeans made a determined effort to settle these lands. The Scots, excluded by English law from the colonization successes in Massachusetts and Virginia, sent thousands of hopefuls to Darién on the Isthmus of Panama, and the French planted a handful of brave souls in Guiana. But the hot, humid tropics generally proved the graveyard of hopes and hopefuls alike.

To effect a wholesale transplant of their ecology, Europeans needed to deploy all of their advantages in an unbeatable combination: their weaponry, their diseases, their animal and vegetable allies, and the evolutionary vulnerability of the environment they were seizing. Outside certain parts of the Americas, they could rely on only their technology and their rapacity, and these were not enough to bring about the domination that they were seeking to impose on the rest of the world.

The Portuguese, who had pioneered the techniques of ocean sailing that opened up the planet, were the first to run up against the limitations of European power. In the late fifteenth and early sixteenth centuries, by virtue of skillful seamanship, they had already found the passage to India that their merchants sought; their well-armed ships could usually be assured of crushing any opposition that they encountered, and they were able to carve themselves a coastal enclave at Goa on the western coast of the subcontinent. But India had a dense population of pathogens—animals and successful crops as well as humans—all of whom had coexisted for millennia. And although the Portuguese managed to hang onto Goa, and indeed earned massive profits from their Indian trade, their soldiers and merchants found the place dangerously unhealthy. There was no prospect of an ecological takeover on the transatlantic model.

The British and French colonization of India also fell far short of complete subjugation. By 1800, the British had built the foundations of a formidable political empire, but although in time they learned sufficient prophylactic medicine to resist Indian diseases, they were never able to refashion their new territorial possession as they had done in America.

The parrot, recipient of the loving gaze of a Dutch woman in the seventeenth-century painting above, was probably brought from the East Indies, one of a number of foreign species to enter ordinary European households as pets following the voyages of discovery and the expansion of trade. For aristocrats, too, for whom the keeping of exotic animals was a long-established tradition of privilege, opportunities widened: The cheetah shown in a detail *(opposite, top)* from a 1765 painting by George Stubbs was the first to be seen in England, a gift to George III from the governor of Madras. Although in all countries the most widely available and popular pets continued to be dogs and cats—such as the kittens *(opposite)* in a detail from an eighteenth-century Chinese painting—the invasion of European domestic space by New World and Asian species prompted a wider awareness of animal types and behavior, and often a new curiosity and affection as well.

It was the same story in the rest of Asia. European sea power granted economic dominion, but it did not guarantee the possibility of settlement in a biologically crowded—and resilient—environment. Although the European trading posts that studded Southeast Asia from the seventeenth century onward often developed into centers of political rule, they never became centers of European population. Even the Philippine Islands, seized by Spaniards who were crossing the Pacific Ocean from Mexico in the mid-sixteenth century, were never Europeanized.

Africa, too, for a long time rebuffed attempts at colonization—defended less by its people, although by that time they had at least learned how to smelt iron, than by its sicknesses. "A striking angel with a flaming sword of deadly fevers," as one explorer put it, guarded the continent against European humans and livestock alike. Although the Portuguese had settlements on West Africa's offshore islands from the middle of

the fifteenth century and the French, the English, and the Dutch all established trading posts during the seventeenth century, the interlopers had to endure appalling losses. Tropical Africa, especially West Africa, was a place where Europeans stayed for brief periods only.

Solely in the far south did Africa offer colonizing Europeans a climate that at least approximated their ideal. The Dutch landed at Cape Town in 1652, and although the cape settlement served mainly as a stopping-off point for ships on the long trade routes to the East Indies, Dutch farmers and their livestock spread slowly northward into open grassland that they called the veld. Unlike the American pampas, though, the African veld was fully provided with grazing animals of its own—as well as human pastoralists, the Bantu, themselves part of a southward movement of African tribes. Dutch muskets guaranteed them victory in most armed encounters, but in general, the density of the local population was too low for it to be destroyed by epidemic disease. The Dutch—and their colonial successors, the English—could dominate, but their progress was painfully slow by the standards of the European advance in the Americas. The Bantu survived, primarily as servants of the white intruders, and their numbers grew faster than their masters'. South Africa never became an authentic transplant from the European stock.

Elsewhere in Africa, most Europeans came only to make a quick profit and then to depart. And in this endeavor they often succeeded, for Africa had one resource that was in great and unceasing demand on the other side of the Atlantic: its people. The massive depopulation of the New World in the early sixteenth century left its Spanish exploiters without a work force for their plantations and silver mines. African slaves were the answer. From 1505, when the first wretched victims were dragged from their homelands to Hispaniola, until the trade finally came to an end in the nineteenth century, at least 10 million Africans crossed the Atlantic in the slave ships' holds. Many more died before they ever saw the New World, either in the prisoner convoys that brought them from the interior of the continent, the coastal barracks where they awaited transport, or in the hideously overcrowded ships themselves.

Black slaves made large-scale sugar production in the West Indies and, later, Brazil not only possible but immensely profitable. The impact of their arrival in the Americas was widespread. In the Caribbean islands, the blacks replaced the original people entirely; in the mines and plantations of Mexico and South America, they lived and bred alongside the surviving indigenes; in the southernmost of the English colonies, they remained a separate field-hand caste.

Apart from irrevocably altering the genetic mix of the Americas' future population, the slave trade had a curious and unintended consequence. In addition to the involuntary passengers in the hold, some of the ships brought stowaways in their water barrels: mosquito larvae carrying with them the plasmodia responsible for yellow fever, malaria, and a number of other insect-borne infections. Once ashore, the insects and their parasites soon found congenial environments in which to multiply, and by 1650, whole regions of South America—Amazonia, for example—that Span-

ish explorers had previously noted as disease-free, became as lethal to Europeans as the West African rain forests. They would remain virtually inviolable, like most of Africa, until the advent of modern medical techniques in the late nineteenth and twentieth centuries made them more hospitable to settlers.

Writing of his voyage to the New World in the 1490s, Christopher Columbus enthusiastically described his discovery as an "earthly paradise." A century later, the English poet John Donne playfully likened his mistress to that still unravished continent: "O my America! my new-found-land." And in the seventeenth century, the philosopher John Locke remarked that "in the beginning, all was America." In fact, neither America nor any other of the lands newly encountered and colonized by European adventurers was virgin territory: They had all been home for thousands of years to other human civilizations, and their inhabitants might well have replied, in the wry words of Miranda in Shakespeare's *The Tempest*, " 'Tis new to thee.' " But to excited and curious Europeans, the freshness of the new discoveries and their apparent lack of recorded history repeatedly evoked images of the Garden of Eden. And just as Adam's first task in the original Paradise had been to name the animals, so the Europeans set about naming, classifying, and ordering the wealth of new life that was set before them.

The desire to capture natural phenomena in the net of language, no less significant a form of taming the wilderness than agriculture or deforestation, was itself far from a new concept. Less legendary ancients than Adam, the most notable of which was

Two views prepared by the English landscape gardener Humphrey Repton in 1790 show *(left)* **his prospective patron's estate as it then existed and Repton's proposal for its aesthetic improvement. The principal changes that were suggested by Repton are the addition of a lake, to be created by damming a river; the conversion of cultivated fields to parkland studded by trees; and the construction of a mock-classical temple, peeping discreetly from the copse on the right. Undulating natural contours, previously abhorred, became fashionable in mid-eighteenth-century England as a backdrop to the mansions of the rural gentry. Accordingly, landscape gardeners rejected the geometrical designs that remained popular in France and employed irregular lakes, terraces, sunken walls, and artfully placed trees as a means of mediating between land used for private recreation and that put to productive use.**

Aristotle, had made determined attempts to find order and pattern in the natural world; and Aristotle's systematic classifications of plants and animals, based on shared and not always obvious characteristics, had at least established that a dolphin was a very different sort of animal from a fish, as was a bat from a bird. But the discovery of the Americas made clear that the ideas and even the knowledge of the ancients, whose revival had helped lead to the Renaissance, were shockingly limited. The world was not only far larger than Aristotle had imagined, it was far more complex, possessed of a wide range of climate and crowded with plants and animals that rivaled anything in the imaginary bestiaries of the Middle Ages. In the peoples that explorers and colonizers encountered, there was an equally baffling diversity of race and culture. God's creation was even greater than had been supposed, his purpose perhaps more inscrutable.

Even without the New World to take into consideration, sixteenth-century classifiers were having trouble. Most of them were medical herbalists—useful plants clearly had a high priority—and they often found it impossible to recognize plants from the descriptions of others, especially if they were northern Europeans trying to decipher a classical work from the different climate of the Mediterranean region. In any case, arranging plants according to their medicinal properties was not likely to cast much light on their natural relationships. There was a desperate need for a system of description and classification—a taxonomy—that not only reflected reality but also would be universally understood.

Observation had to precede description, and for this purpose, it was necessary to gather together representative specimens of the newly found flora and fauna. The first botanical garden in Europe was established at Padua in 1545, partly to observe new plants brought back from abroad by the navigators, and other such gardens were soon flourishing elsewhere in Europe. In the course of the following two centuries, English, Dutch, French, and Spanish gardens were planted with exotic trees and other imported species, and these were complemented by zoos, greenhouses, and museums for the study of animals, plants, seeds, and rocks. George Stubbs, the celebrated English painter of racehorses, took time off from his equine routine to paint the first zebra and the first cheetah to be brought to England—from South Africa in 1762 and from India in 1764, respectively.

By the seventeenth century, empirical observation and intellectual developments in many different fields had fermented together into what was later known as the scientific revolution, not the least because of the invention of a number of new instruments—among them the microscope, the telescope, the thermometer, and the barometer—that shed light on phenomena invisible to the naked eye. The study of physics was the greatest beneficiary of the newfound knowledge, but from the herbals and bestiaries of the past, the new sciences of botany and zoology were quickly emerging. Their practitioners, with additional information at their disposal and the availability of advanced printing techniques to disseminate it widely, were learning to classify their increasing number of specimens into groupings that shared multiple and complex characteristics.

In the eighteenth century, Swedish botanist Carl von Linné produced a system that sacrificed some of the natural categories of earlier rivals for ease of use and consistency: Genera and classes of life were established according to simple and often countable criteria. Linné also introduced the system of scientific nomenclature that has survived to the present day, virtually inventing a new dialect of Latin for the purpose, and it is in the Latin form of Carolus Linnaeus that his own name has been remembered. The Linnaean system of classification allowed explorers to easily describe what they had seen, and a torrent of specimens, drawings, and notes flowed back to the universities, scientific societies, and private studies of Europe. Meanwhile, the French naturalist Georges-Louis de Buffon was beginning work on his encyclopedic *Histoire Naturelle,* of which he managed to complete just thirty-six out of a proposed fifty volumes before his death in 1788. Buffon collated a wide range of previously unconnected data and proposed a revolutionary theory of geological history that was to be further developed by Charles Darwin.

The overall effect of the new intellectual developments was to enshrine scientific knowledge and physical order as the gods of European civilization. The people of Europe also acquired a new confidence in their own authority over the natural world. The French philosopher René Descartes had defined man as a thinking being—animals, on the other hand, were mere robots, incapable of either thinking or feeling pain. Scientists such as Galileo and Isaac Newton had asserted that nature obeyed mechanical laws and that human beings had the power to predict events in the

In a painting by the English artist Joseph Wright exhibited in 1766 *(left)*, a philosopher makes use of an orrery—a mechanical model of the solar system—to explain the theories about the operations of the universe proposed by astronomers and scientists such as Copernicus, Kepler, and Newton. By turning a handle, the philosopher could cause the planets to rotate on overlapping concentric bands around an object representing the sun, in this case a lamp wick burning in oil. Such public demonstrations made known to a wide lay audience—which here includes a woman and three children—scientific knowledge that, as the French writer François Voltaire noted in the 1720s, had previously been much talked about but only rarely understood.

natural world. To those fully converted to the new doctrines, moreover, the application of science and human reason appeared to promise freedom from want, scarcity, and the arbitrariness of natural calamities.

Within Europe, where the population began to increase rapidly around the middle of the seventeenth century—partly as a result of dietary improvements derived from New World plants such as corn and the potato—human ingenuity and muscle power were exploited in an effort to increase the production of food. Vast projects of land reclamation and drainage were undertaken in northern Italy, western France, the Netherlands, and the fens of eastern England. Deforestation in northern Russia cleared new land for agriculture, and the untamed wilderness of Siberia gradually yielded to human settlers. Applied to stockbreeding in western Europe, scientific knowledge aided the development of new breeds of cattle, sheep, and pigs that yielded more fat and meat than their forebears. In Britain, the invention of new agricultural machinery—such as Jethro Tull's mechanical sower, which planted seeds in neat rows—enabled the labor of planting and harvesting crops to be carried out by fewer people in less time. Such technological developments were harbingers of the Industrial Revolution, which in the nineteenth century would permit human beings to exploit natural resources with greater efficiency than the most anthropocentric naturalist had ever imagined possible.

Outside Europe, too, science was altering attitudes toward nature and providing the impetus for further development. The explorers of the second great age of European discovery in the eighteenth century, unlike their fifteenth-century equivalents, were usually naval officers who were sailing on the orders of their government in well-equipped ships carrying scientific observers and sophisticated instruments for measuring and recording. When the Royal Navy's captain James Cook, the most painstaking navigator of his time, made landfall on Australia's habitable eastern coast in 1770, he was accompanied by the eminent botanist Joseph Banks, and their primary concern was to record and communicate hard scientific data about their discovery. Cook's own sympathetic description of Australia's Aborigines—"They may appear to some to be the most wretched people on earth," he wrote, "but in reality they are far happier than we Europeans"—reflected a new attitude toward the people of other cultures, which was based more on curiosity and interest than on the unrestrained will to dominate them.

In the end, however, scientific objectivity was not enough to blunt the ecological impact of the Europeans on a continent that had been separated from the main planetary landmass for even longer than the Americas. The Aborigines of Australia, hunter-gatherers with no weapon more powerful than a stone-tipped spear, no domestic animals other than hunting dogs, and no epidemic diseases whatsoever, were even less capable of defending themselves and their environment than the native Americans had been. Following the establishment of a British penal colony at Sydney Cove in 1788, the colonization of Australia proved to be a recapitulation, on a compressed timetable, of the American experience almost three centuries earlier. The colony's governor went out of his way to avoid harming the original inhabitants, but his conscientiousness did no good. By 1789, Aborigines who had come in contact with the newcomers were dying of something akin to smallpox, even though the long ocean voyage of the convict settlers should have served as an effective quarantine. Passed from band to band, the disease swept inland along the hunting routes, killing perhaps one-third of Australia's indigenes before the English settlers had even moved out of sight of the sea.

As in America, European livestock continued the work of conquest. The eight beasts that had been purchased by the British fleet of 1788 at a Cape Town stopover escaped into the hinterland. By 1804, the strays had bred into a herd of 5,000 wild cattle, at once a resource for the expanding colony and a biological bulldozer that opened niches for European plants and people as it spread and reproduced. The Aborigines, whose hunter-gatherer culture demanded vast spaces for every family, would be squeezed toward extinction. Within a very few human generations, the cattle would be joined by millions of sheep, their wool destined to supply Britain's newly mechanized textile industry. Australia would become a European-style state with an English-speaking population millions strong.

By 1800, the expanding European ecology was nearing its limits. Of all the temperate lands whose inhabitants might conveniently be dispossessed, only New Zealand,

The knowledge won by scientific advances in the seventeenth century increased the human capacity to modify and—as the advocates of progress insisted—improve the given shape of the natural world. Challenged by the demands of populations that in Europe began to rise steeply around 1650, agriculturalists in Britain and the Netherlands especially put this power to use. Landholdings were amalgamated and enclosed to create compact farms that were economical to operate; estate owners experimented with new crops and systems of crop rotation and invested in mechanical sowers and factory-made plows; the size of livestock was increased by scientific methods of breeding; and new land was claimed for agriculture by extensive drainage. The panorama shown on the right and extending overleaf shows an estate in Gloucestershire in 1730: Behind teams of laborers gathering hay into shocks and advancing in line with scythes, more distant fields are patterned by the regular ridges and furrows left by the plow. In England alone, 6.2 million acres were enclosed and worked in this manner between 1700 and 1845.

IMPROVING ON NATURE

mapped by Cook but devoid of white settlement until 1814, remained. The great if somewhat one-sided exchange of plants and animals that had begun in the time of Columbus was almost complete: Even before the arrival of alien settlers, New Zealand's Maoris were noting with consternation the strange new weeds that sprang up on their islands. Potatoes were fast becoming a dietary staple in Europe itself; corn and cassava were firmly established in Africa. The European hegemony over the seas had been responsible for the transfer of other plants, too: Breadfruit, for example—from Tahiti and Hawaii, whose populations were now suffering from their first experience of epidemic disease—had been naturalized in the West Indies as a cheap nourishment for plantation slaves.

But while the ecological explosion detonated by the impact of European colonization on the wider world began to subside, the narrative of human adventure was still a long way from any satisfactory conclusion. The domination of a previously localized, Europe-centered civilization over a large part of the globe's landmasses was now ensured, and it had been achieved in a remarkably short period of time; but

the new overlords' comprehension of the land they laid claim to lagged far behind their actual expansion.

Emblematic of this failure in understanding was the experience of the expedition led by Meriwether Lewis and William Clark, which set out in May 1804 from Saint Louis to reach the western edge of the American continent. Comparable in ambition to Columbus's transatlantic venture three centuries earlier, the Lewis and Clark expedition, sponsored by President Thomas Jefferson, started out with the most up-to-date equipment available: scientific instruments, vaccines, friction matches, textbooks on botany and mineralogy, an air gun, a collapsible boat. They were imbued, too, with the most enlightened ideas of the time, instructed by Jefferson to treat the Indians well and to record details of climate and of flora and fauna with "great pains and accuracy." Even so, they were wholly unprepared for the landscape they traversed, which seemed to mock their puny baggage. Assailed by storms, cacti, mosquitoes, rattlesnakes, cougars, and grizzly bears and compelled to survive off melted snow and the flesh of their own horses, Lewis was driven to report that "to

In a painting dated 1810 (above), the English stockbreeder Robert Bakewell hires out his rams to farmers seeking to improve their own herds. By methodical selection and culling, Bakewell developed early-maturing sheep with a high yield of meat.

make any further experiments in our present situation seemed to me madness." The expedition did eventually reach the Pacific Ocean, but for his remaining years, Lewis was recurrently subject to "sensible depressions of the mind." He died in 1809 of gunshot wounds, probably self-inflicted.

But if the European expansion had proceeded at a pace that exceeded understanding, it was now too late to slow it down. Industrial development and the growing population of Europe accelerated the momentum, and in the ensuing decades of the nineteenth century, the route of the Lewis and Clark expedition was to be followed by wagon trains bearing thousands of hopeful settlers. After overcoming their preconception that only land where trees grew in abundance was properly fertile—a belief that had led previously to massive deforestation—and aided by the steel plow, barbed wire, repeating rifles, and advancing railroads, newcomers from Europe turned the Indian hunting grounds of the Great Plains into a profitable granary and marched westward across the Rocky Mountains in quest of California's gold. They were just one contingent of fifty million emigrants who crossed the oceans from Europe during the century, most of them to the United States but many to the pampas of South America and to Australasia. These areas had a total population of little more than five million as the century began, far less than the 12 to 15 million of the pre-Columban Americas alone; natural increase combined with immigration made it close to 200 million in 1910.

By then, almost 90 percent of the population of the United States was of European descent, with Africans accounting for most of the rest and the descendants of the original Amerindians making up only a tiny fraction. But numbers alone were only one indication of the changes that had taken place. Even more significant, the uses to which the land was put were now determined by cultural attitudes wholly different from those of its original inhabitants. "We did not think of the great open plains, the beautiful rolling hills, and winding streams with tangled growth as wild," declared a Sioux Indian chief. "Only to the white man was nature a wilderness. To us it was tame. Earth was bountiful, and we were surrounded with the blessings of the Great Mystery." That mystery was being eroded.

THE ODYSSEYS OF PLANTS

Plants provide food, shelter, fuel, clothing, and medicine; they are also used to make furniture, paper, pigments, cosmetics, and photographic film. Among the most precious resources of any society, they have always excited the envy or greed of those who do not possess them, and their dissemination from one part of the globe to another has been speeded by political no less than by natural currents.

All of the plants that appear here and on the following pages now grow in countries to which they were taken by human agents, often as trade goods or in the wake of military conquests. Their movement around the world has altered landscapes and revolutionized economies. The coffee plant, shown on the right in a seventeenth-century French engraving, for example, originally grew wild in Ethiopia, from where it was taken by traders to the Arabian Peninsula in the fifteenth century. Coffee beans were imported into Europe along with a range of spices from the Far East, and the popularity of the drink prompted the colonial powers in the seventeenth and eighteenth centuries to cultivate the plant in their overseas possessions. A single bush was carried to Martinique in 1723; from the Caribbean, coffee plantations spread to coastal South America and Brazil, which is now the largest coffee producer in the world.

The coffee plant and the other plants shown here are only a sampling of those spread from their original habitat by humans. Food plants alone include the tomato, the banana, and the chili. And there are undoubtedly others that still await their dissemination. Studies of tropical rainforest plants conducted in the 1980s, for example, found that many of them contain chemical compounds previously unknown

A sixteenth-century engraving *(right)* shows a sugar mill in Sicily: Juice from sugarcane is pressed out, boiled, and poured into conical molds to set. Sugarcane, depicted below in an illustration from a 1615 treatise, originated in India and was introduced to Europe after the Muslim conquests of the seventh century. Christopher Columbus carried seedlings from the Canary Islands to Hispaniola in 1493; within a few decades, most of Europe's sugar came from the West Indies.

A Peruvian pot *(far left)* dating from the tenth century is shaped in the form of two potatoes, which grew wild in South America as early as 6000 BC. Carried to Spain in the 1530s by the conquistadors, the potato at first fetched high prices as a cure for impotence. By the late eighteenth century, however, it became a common European food. The failure of potato crops in Ireland in 1845 caused widespread famine, illustrated by this nineteenth-century engraving of women and children searching for potatoes that had escaped the blight; starvation and emigration reduced Ireland's population by nearly one-fourth in the next six years.

Dating from the fourteenth century, a life-size corncob with silver grains and husks *(far right)* testifies to the plant's importance to the Incas of Peru. The dietary mainstay of all early Central and South American peoples, corn was first cultivated about 5000 BC. The plant was taken from the Caribbean to Spain by Columbus in 1496; a century later, it was planted throughout Europe and in parts of the Middle East. On the right, a detail from an eighteenth-century Venetian painting shows a woman turning out a pot of polenta, made from yellow cornmeal, which became a staple food of the poor in northern Italy.

A picture of the peanut plant *(left)* from a book published in the Philippines in the nineteenth century shows its unique pattern of growth: After fertilization, the flowers burrow underground, where the seeds ripen in brittle shells. Native to tropical South America, the plant was taken to Europe by Spanish explorers and then to countries in Africa and Asia. Brought to North America as a food for slaves, it became popular during the Civil War, when Union soldiers—as shown in the painting by Winslow Homer *(below)*—roasted the nuts on their campfires. Except in the United States, the plant is grown mainly for its oil.

An Egyptian tomb painting dating from 1200 BC *(below)* shows the harvesting of flax, whose fiber was used to make linen for clothes and to wrap the bodies of the dead. Flax was also grown in prehistoric Europe, but not until the Romans introduced linen to France and Belgium was it used to make clothing on a large scale.

On the right, an illustration from a fifteenth-century Italian edition of the work of the Roman natural-science historian Pliny the Elder shows women separating flax fibers from the woody stems. Linen was the most common clothing material in Europe until the advent of inexpensive cotton in the nineteenth century.

In a game recorded by a French artist in 1591 *(right)*, Central American Indians bounce rubber balls made from the milky sap of a native tree. After noting local products derived from another species of the tree native to the Amazon basin, Europeans began to use rubber in the early nineteenth century. Plants grown at Kew Gardens in London from seeds brought from Brazil were dispatched in the 1870s to Ceylon and the Malay Peninsula, where plantations were established. Henry Ridley *(opposite)*, director of the botanic gardens in Singapore from 1888 to 1912, demonstrates the rubber-tapping method that he devised: a herringbone pattern of incisions that ensures an easily collectable flow of latex while allowing the bark to renew itself smoothly.

Fabric made from the cotton plant—depicted on the right in a nineteenth-century Indian drawing—has been found in tombs in India and Peru dating from before 2000 BC. The plant later became widely dispersed: It grew around the Mediterranean in Roman times, reached China in AD 600, and was introduced to Africa by the Arabs in the eighth century. Cultivation in North America began early in the seventeenth century; the labor of African slaves, shown in the nineteenth-century engraving above, and the invention in 1793 of the cotton gin made the United States the world's largest producer.

In an early-nineteenth-century watercolor *(above)*, Dutch merchants sample tea in a Chinese warehouse. Tea was drunk in China as early as 2700 BC. By the eighth century, tea plants were growing in Japan and other Asian countries. Imported from Java by the Dutch in 1610, tea became popular in Holland, England, and the American colonies. The British found tea growing wild in India and began to cultivate it there in 1834; by 1900, more than 760 plantations covered the hills of Assam *(right)*.

The tobacco plant—shown in the sixteenth-century Dutch engraving on the far left, the first published picture of the plant—took its name from an American Indian word for a leaf cylinder prepared for smoking or a tube for inhaling the smoke. In the 1550s, the plant, native to tropical regions of the Americas, was taken to Europe, where it was at first prized as a panacea for everything from snakebites to ulcers. Smoking was popularized in England by Sir Walter Raleigh; most Europeans smoked pipes, as shown in an illustration *(near left)* from a seventeenth-century Dutch pamphlet, but Spaniards preferred cigars. Cultivated by North American settlers from 1612, tobacco became the main commodity traded by the colonists for European goods. Tobacco is now grown in about eighty countries.

In a nineteenth-century engraving, bark is stripped from cinchona trees in the forests of the South American Andes, the plant's native habitat. The protection against malaria afforded by quinine, a drug derived from the bark, was first made known in Europe by Jesuit missionaries returning from South America in the seventeenth century. The Dutch and the British later grew their own supplies in Indonesia, Africa, and India, where quinine facilitated their conquests and was sold in packets *(inset)* to colonial officials.

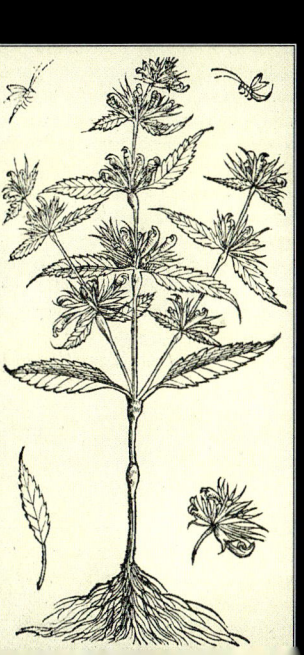

PURE QUININE.

5 grains one dose.

Price 1 pice.

A fourteenth-century Persian miniature *(far left)* shows a woman enjoying the effects of cannabis. The plant, seen at left in a seventeenth-century French drawing, originated in central Asia and was grown in China as early as 2700 BC. It was cultivated in Europe during the Middle Ages, in Chile from the early sixteenth century, and in British colonies in North America a century later. Also known as Indian hemp, it is grown for its tough fibers—used to make cloth and rope—as well as for the drug derived from its resin. Cannabis was once used as a sedative and analgesic but has no accepted use in modern medicine.

THE GLOBAL CHALLENGE

One day during the 1930s, a motorist was driving on a country road across the Great Plains of the United States. Seeing a ten-gallon hat resting on a nearby sand dune, he stopped his car and got out. Under the hat, so the story goes, he was surprised to discover a man—or at least the head of a man. "Looks as though you're in trouble," said the traveler. "Let me give you a lift into town." "I'll get there myself, thanks," replied the head from its pile of sand. "I'm on a horse."

The situation behind this story was in fact the worst ecological disaster that the United States had experienced, a disaster of such magnitude that rueful humor was the only way many Americans could cope with it. Overcultivation was part of the cause. Posters during World War I had urged: "Plant more wheat—wheat will win the war." The farmers of the southern Great Plains had done so; and in the postwar years, they continued to plow up the fragile grasslands of Kansas, Oklahoma, and northern Texas. Then, in the early 1930s, the rains stopped, and the land grew bone-dry. With nothing to anchor it, the precious topsoil swirled across the Plains, bringing agriculture to a standstill and sending thousands of impoverished farmers on a pitiful exodus to California.

The blame for what had happened was difficult to apportion. The farmers pointed to the politicians, and the politicians pointed to the American public, whose appetite for inexpensive bread, they claimed, encouraged the sort of agriculture that could transform a sea of living grass into a sterile desert. But in a sense, the responsibility was even more widespread, for the tragedy of the Dust Bowl—as this region became known—was just one of many painful lessons that humanity throughout the world was being forced to learn.

Environmental crises were not a new phenomenon: Erosion, pollution, and the extinction of species had for centuries tarnished the achievements of civilizations. During the nineteenth century, however, the scale of humanity's impact on the planet began to assume menacing dimensions. Industrialization, technological developments, and an enormous growth in the world's population confronted people everywhere with sobering evidence of environmental deterioration, ranging from lifeless waters and barren hillsides to the asphyxiating air of smoke-choked cities. Voices were raised in protest, but it was hard to argue against the powerful assumptions of progress and growth that governed economic and political thought, and industrial civilization hurtled into the twentieth century with a reckless momentum that threatened ever-increasing catastrophes. Margins for error were progressively reduced. A simple accident could turn an ocean black with oil or fill the air with a deadly cloud of poisonous vapor. Nuclear weapons promised destruction on an inconceivable scale. A disturbing new lexicon of ecological concepts entered the vocabularies of ordinary men and women: acid rain, ozone depletion, desertification, nuclear fallout,

A helicopter flies above the devastated flanks of Mount Saint Helens in Washington State, spraying fertilizer on grass seeded to limit erosion after a massive volcanic eruption on May 18, 1980. The eruption tore more than 1,300 feet off the summit of the 9,677-foot-high volcano, destroyed tree cover over an area of about 200 square miles, and claimed the lives of about sixty people. "What we have here," declared a biologist surveying the gray wasteland, "is almost a Genesis situation." As such, it demonstrated nature's own regenerative powers. The deep roots of fireweed plants sent up shoots that broke the ash-encrusted surface; browsing animals and insects feeding on rotting debris gained a foothold; and rains restored the soil's fertility.

An 1837 lithograph shows laborers at work at the mouth of a tunnel on the railroad line between London and Birmingham, which was opened in the following year. Unlike highways, railroads had to be built as straight and as level as possible: Experiments made in 1833 showed that a train ascending a gradient as slight as 1 in 300 needed twice the tractive power required to haul the same load over flat ground. Therefore, hills were leveled or pierced by tunnels, and valleys were filled in or spanned by soaring viaducts. Some 20,000 men labored for five years to construct the 124-mile London-to-Birmingham line, a task compared to the building of the pyramids in ancient Egypt.

global warming. And as the century drew to a close, the earth's ability to sustain human life under all circumstances—a premise past civilizations had never doubted—was increasingly seen as a dangerous illusion.

About 1650, there began a sustained and rapid surge in the human population that was to alter the face of the globe. At that time, the earth was home to around 500 million people—the number had doubled roughly every 1,500 years since 8000 BC. Over the next two centuries, the population doubled again, reaching one billion by 1850. Most of this growth was centered in Europe, whose population rose from 140 million to 266 million between 1750 and 1850. The inhabitants of Britain and Ireland tripled in number over this period, in spite of waves of emigrants who fled to the more open spaces of America.

This rapid increase was initiated by improvements in agriculture and a growing spirit of free enterprise. Since they were no longer bound to the land as peasant farmers, Europe's rural masses flocked to the cities, ambitious for work and wealth. Many died there in wretched poverty, but the sustaining glimmer of riches seemed always within their reach, and its flame was fanned by tales of wealth from beyond the seas. Indeed, there was soon evidence of the New World's bounty, not just in the gold and silver—or codfish and pelts—that flowed east across the Atlantic Ocean but in the new food crops, particularly potatoes and corn, that in turn helped support even greater numbers of people.

In the foreground of *Mountain Landscape with Rainbow,* painted by the German artist Caspar David Friedrich around 1810 *(above),* a solitary spectator stands in rapt contemplation of the majesty of nature. Artists and writers of the Romantic Movement gave definitive expression to new attitudes toward the natural world that had been evolving since the voyages of exploration and the scientific discoveries of Copernicus, Galileo, and Newton in preceding centuries: The immensities of nature itself now rivaled God as the proper object of human awe. Like high-class tourist brochures, canvases such as that shown here prompted the leisured classes to make excursions to the Alps, the Scottish Highlands, or the English Lake District, their journeys facilitated by Europe's expanding network of railroads.

As Europe's population expanded, more land was required for crops. The threat to the forests of England was particularly pressing. Substantial stands of timber had to be preserved for the maintenance of the navy, and wood was also the prime fuel for industry and domestic heat. The only alternative was coal—an organic mineral rich in carbon—and in this resource England was extremely wealthy. The scale of mining had been steadily increasing since medieval times, and by the end of the seventeenth century, most of English industry, from brewing to brickmaking, was fired by coal; only iron smelting remained dependent on charcoal, derived from the burning of wood, and this last bastion of traditional industry toppled in 1709 when Abraham Darby developed a technique for using coke to fire his furnaces in Shropshire. Although the full impact of these developments did not become apparent for more than a century, the Industrial Revolution was under way. It would alter forever humanity's relationship with the natural world.

The mining and transportation of coal tested the ingenuity of English inventors and engineers. Shallow open-cut pits were quickly exhausted, and because deep mines tended to fill with water, the need arose for an efficient pump. The steam engine was first developed for this purpose but soon proved to have a wide range of other applications. Coal had to be carried in bulk from the mines to the centers of industry, and the use of horses to pull both wagons and canal barges was inefficient and wasteful. "It has been said that in Great Britain there are above a million horses engaged in various ways in the transport of passengers and goods," reported a select

A pair of flightless great auks are captured in characteristic poses in this hand-colored engraving from *Birds of America*, a pictorial survey published between 1827 and 1838 in which the artist James Audubon sought to record every living bird species in North America. Great auks, which for centuries had been protected by the isolation of their habitat along the coast of Newfoundland, were killed in increasing numbers after the arrival in the sixteenth century of European fishing fleets, which provisioned their return voyage across the Atlantic with the birds' meat and eggs. Little more than a decade after Audubon sketched these and other birds on a trip to Labrador in 1833, the great auk was extinct.

committee of the House of Commons in 1833, "and that to support each horse requires as much land as would upon an average support eight men." The solution lay in steam-powered trains, first introduced in 1825. By the mid-nineteenth century, Britain was bound in a web of iron rails.

The revolution took on a momentum of its own, one invention igniting another like explosions on a string of firecrackers. In the 1850s, the English engineer Henry Bessemer developed an economical means of converting pig iron into steel. By 1859, the United States had begun to tap its underground reservoirs of oil, the fuel that was destined to transform yet again the development of transportation and industry. Innovations in the electrical and chemical industries opened even more doors in the last quarter of the century.

The harm caused to the natural world was soon apparent. Industrial waste and untreated sewage poisoned rivers and canals. Factory chimneys spewed out sulfur dioxide, produced by the burning of coal, which defoliated the surrounding countryside. Cheap housing for the armies of new factory workers sprawled across former farmland. Villages that had scarcely changed in the previous century became lost overnight in a labyrinth of narrow streets and sooty brick. Farther afield, valleys were flooded to serve as reservoirs for the new conurbations, while the spreading railroads opened up new regions for development.

Although France was rapidly following England's lead in methods of production, the historian Alexis de Tocqueville was unprepared for the sheer ugliness of Britain's industrial heartland. "Thirty or forty factories rise on the top of the hills," he wrote of Manchester in his *Journey to England and Ireland,* published in 1835. "The wretched dwellings of the poor are scattered haphazardly across them. Around them stretches land uncultivated but without the charm of rustic nature, and still without the amenities of a town. The fetid, muddy waters, stained with a thousand colors by the factories they pass, wander slowly around this refuge of poverty." Not everyone shared de Tocqueville's dismay. To many nineteenth-century Europeans and Americans, the Industrial Revolution marked an important advance in human civilization.

There was something noble in the power that human beings could now harness for their own enrichment. "Examine the endless varieties of machinery that man has created," wrote one English enthusiast in 1831. "Mark how all the complicated movements cooperate, in beautiful concert, to produce the desired result.... We do entertain an unfaltering belief in the permanent and continued improvement of the human race, and we consider no small part of it, whether in relation to the body or the mind, as the result of mechanical invention."

Such sentiments had global consequences. France, Germany, and the United States were early converts to the doctrines of industrialization, and Russia and Japan followed later in the century. And because Europe could not supply its voracious new industries or expanding population solely from its own resources, goods from abroad became essential buttresses of the Old World's economy: furs from Canada, cotton

This sperm whale's tooth from the Pacific Ocean was carved with a picture of a whale being harpooned during the mid-nineteenth century, when America's whaling fleet numbered more than 700 vessels. Such artifacts were produced by crewmen during their long periods of idleness as they sailed to the whaling grounds or waited for their prey to blow. Unlike the Inuit of eastern and western North America, who exploited every part of a whale's carcass for food, fuel, clothing, and tools, American hunters were interested only in the whales' blubber, a major source of oil for lighting and heating before the discovery of petroleum in 1859.

from the American South and India, beef from the Argentine, wool from Australia, hides, silk, jute, and rubber. In many cases, these goods answered a genuine need: Because sheep consumed valuable farmland, for example, England's long-established wool industry was forced to contract, and by the middle of the nineteenth century, working men and women were clothed almost entirely in imported cotton cloth. But with rising standards of living, need accumulated on need, playing havoc with the economies and the environment of the non-European world. In addition, European capitalists were committed to opening up new markets abroad—and the same ethic that permitted the exploitation of both labor and land in Europe encouraged them to look on the undeveloped world as little more than a giant factory, fit only for the production of food and supply of resources.

The traders and entrepreneurs who traveled outside Europe were not immune to the wonders of their new surroundings. "The larger portion of this desolate region is covered with lofty and luxuriant green forests in every direction," reported two British engineers who explored India's mountainous southwestern region in 1817. "The whole scene is truly sublime." But profit took precedence over beauty, and by the end of the century, tea and coffee plantations covered a large part of the mountain areas. In Assam, at the opposite corner of the subcontinent, 764 plantations were exporting about 145 million pounds of tea annually by 1900, while an army of 400,000 low-caste workers, living in virtual slavery, ensured that the teapots of Britain were never empty.

The transformation of agriculture in the undeveloped nations was not always the result of European occupation. Even before colonial rule was established in West Africa, for instance, native farmers had acknowledged the commercial advantage of growing cash crops for Europe. A swift steamship service, initiated in the 1860s, carried their produce northward: peanuts, palm oil, and cocoa. And the practice of growing crops for sale to the industrialized world, to the neglect of essential food crops at home, continued through colonial days and into independence, sometimes at a severe cost to the environment. Silt from deforested hillsides clogged lowland rivers and watercourses flooded with destructive regularity, while foraging cattle ensured that a new tree cover would never establish itself naturally. "All progress in capitalistic agriculture," Karl Marx grimly observed, "is a progress in the art, not just of robbing the laborer, but of robbing the soil."

Other agents besides humans had a devastating effect. Wherever a ship sailed into an undiscovered bay or a convoy of wagons lumbered across a trackless plain, there advanced an army of living things of which human beings constituted only a minority. Rabbits were introduced to Australia in 1859; within a century their numbers had increased to an estimated 500 million. Pigs, dogs, cats, goats, and the adaptable honeybee likewise thrived in the new worlds to which they were transported. And wherever European men and women traveled, that most faithful of all domestic animals—the rat—was at their side.

Vermin and overgrazing by European livestock created conspicuous environmental problems in newly explored countries. Native plants were forced into retreat, often to the point of extinction. Thirty-three plants unique to the island of Saint Helena

Two white overseers *(above)* lounge by the side of a path leading through a banana plantation in Angola around 1890. To alleviate the financial burden of administering their African colonies in the late nineteenth century, European powers encouraged the growing of cash crops and laid down a transportation network to speed their export, thereby subjecting the land to the same spirit of capitalist enterprise that had already left its mark in Europe. The railroad shown on the left, flanked by some of the 32,000 Indian workers who were imported to construct it between 1896 and 1901, was laid over some 620 miles of difficult terrain between Mombasa on the coast of Kenya and the northeast shore of Lake Victoria in Uganda.

were identified in 1810; since then, goats have eliminated all but eleven of them. Native animals were also threatened. Rats and pigs ate the eggs and young of ground-nesting birds. The dodo, native to the islands of Mauritius and Réunion, became extinct by the late seventeenth century. The notornis, a similar flightless bird of New Zealand, was unable to compete for grazing with the European red deer and barely avoided the same fate. In Australia, an expedition in the Murray-Darling region northwest of Melbourne discovered thirty-one species of mammals in 1856 and 1857; only nine of these survive.

A quieter but no less remarkable revolution was occurring in the vegetation of the colonized lands. Some plants arrived as official immigrants: More than 200 species were deliberately introduced into Australia by 1803. Hundreds more arrived as stowaways, their seeds mixed in with others or clinging to the coats of animals. Their spectacular success literally changed the face of much of the world. Only one-fourth of the plants growing on the Argentine pampas today are native species. Of the principal weeds in North America, more than half are of Old World origin. Only in the tropics did native species hold out successfully against the aggressive immigrants.

The most unwholesome allies of the European colonists were their own pathogens. Smallpox was less widespread in the nineteenth than in earlier centuries, but the invaders still maintained a formidable arsenal of germs. "Wherever the European had trod, death seems to pursue the aboriginal," observed the English naturalist Charles Darwin in 1839. The Maoris of New Zealand, a determined and populous warrior people, fell by the thousands to tuberculosis and measles, while venereal disease introduced from the Continent kept their birthrates artificially low. By the end of the century, their numbers had been reduced to around 42,000, from the nearly 200,000 that had occupied the islands on Captain Cook's arrival in 1769. A melancholy proverb circulating among the Maoris in the mid-nineteenth century summed up the

Goats and cattle forage *(below)* along the shifting border between desert and pasture in Niger, West Africa. On the right, a wood vendor travels toward a local market in Senegal, where timber is the only fuel available for nine out of ten rural households. Overgrazing and tree cutting, which can render fertile land barren, contribute to the encroachment of desert on the arid but potentially productive territories that cover more than one-third of the world's land surface.

problem facing aboriginal people throughout the temperate world: "As the white man's rat has driven away the native rat, so the European fly drives away our own, and the clover kills our fern, so will the Maoris die before the white man himself."

"I wish to speak a word for nature, for absolute freedom and wildness," announced Henry David Thoreau to a gathering in his native Concord, Massachusetts, in 1851. It was a subject with which the young man was well acquainted. A keen naturalist and woodsman, he had emerged four years earlier from a solitary life in a log cabin on the shores of Walden Pond, "living deep and sucking all the marrow out of life," as he described the experience. Thoreau's self-appointed mission was to persuade his fellow Americans that nature was a source of inspiration without which human existence was superficial and empty. "In wilderness is the preservation of the world," was the rousing climax of his speech in Concord that day.

Thoreau's tone of impassioned urgency was appropriate, for in his own nation, the wilderness was succumbing to industrial and material progress on a greater scale than anywhere else. Alexis de Tocqueville, visiting the United States in 1831, was astonished by the brutality with which Americans treated their natural surroundings. "They are insensible to the wonders of inanimate nature," he complained, "and they may be said not to perceive the mighty forests that surround them till they fall beneath the hatchet." In the 1840s, almost two million acres of redwoods, the earth's tallest living things, were growing in California; by the time their systematic felling was finally halted in the twentieth century, 96 percent of the mature trees had vanished.

Trees were for felling, animals and birds for shooting—and the rifles that slaughtered boars by the thousands on the great hunting estates of northern Europe had an even more dramatic effect in America. In the early nineteenth century, no American bird was more numerous than the passenger pigeon. Hundreds of millions of these creatures darkened the skies in flocks that might take two or three days to pass overhead. Early frontiersmen and farmers shot them for the cooking pot, but with a hungry new market in the eastern cities, the culling of pigeons became a massacre. The birds were caught in nets, drugged with alcohol-soaked corn, asphyxiated with smoke, or simply blasted out of the sky. Their bodies were loaded into freight cars and hauled back to the markets of Philadelphia or New York. In 1878, one hunter in Michigan boasted an annual bag of three million birds. Within a decade, their numbers had declined below a sustainable level. The last of the species died in a zoo in Cincinnati in 1914.

A similar story was unfolding on the Great Plains. An estimated 30 to 40 million American bison—or buffalo, as they were popularly called—ranged the short grasslands of America's western territory at the beginning of the nineteenth century. Even equipped with rifles and mounted on horseback, the Plains Indians who hunted them

Three watercolors from a 1608 treatise by Guiseppe Paulini, an Italian landowner, show the effects of deforestation on a hill near Venice. After felling the trees, farmers fired the stumps to make fields; the topsoil, no longer anchored by the trees' roots, was washed away by rain, leaving the hillside bare.

This view of the Grand Canyon, painted in 1872, helped persuade the U.S. government to create Yellowstone National Park.

In the nineteenth century, human impact on the natural world was suddenly speeded up: Vistas of untrammeled wilderness and of plains and valleys that had for centuries been marked only by the farmer's plow now increasingly bore the scars of industrial development. For many people, this acceleration prompted a late concern for what was being lost, and a commitment to preserve what was left.

For some, the value of untamed nature was perceived as spiritual: Without wilderness, they claimed, the human soul would be diminished. For others, nature's worth lay in the timber, soil, and minerals that supplied the requirements of civilization. The lesson that these resources were finite had been learned and relearned by every urban society since the Sumerians in Mesopotamia in 3000 BC; and the ruinous ways of humans themselves were also well known, as illustrated in the images on the far left of a denuded hillside. The challenge that united idealists and materialists was how to protect the world both from and for its dominant species.

The world's first national park was established in California in 1864; by the 1920s, such parks existed on every continent. The preservation of forests and grassland within these regions maintains soil productivity and the supply of water for urban reservoirs. Parks also have an aesthetic and recreational value, and often yield large revenues from tourism. But probably their most important benefit is the protection they afford to a still unknown number of plants and animals whose life cycles and food chains are essential to the existence of all forms of life, including humans.

Elephants range across the grassland of Tsavo National Park in Kenya, prey for tourists' cameras rather than poachers' guns.

had made no significant impact on their numbers. Nor did they want to, for their way of life depended on the buffalo's survival. To whites, however, the buffalo was a crop to be harvested. The slaughter began in earnest in the 1860s and reached its height between 1870 and 1875, when an estimated 2.5 million buffalo were killed each year. Their hides were baled up and sent off to tanneries; their bones were shipped east to be ground up for fertilizer. By the end of the century, there were perhaps 500 surviving Plains buffalo in the United States.

Against destruction on such a monumental scale, Thoreau's angry cry of protest could by itself achieve little. Fortunately, he was not alone. In Europe at the beginning of the century, a number of artists and writers, repelled by the squalor and filth of the cities, had begun to look to the natural world not just as a material resource but as the font of spiritual wisdom. They became known as the Romantics, members of a cultural movement that caused a significant change in attitude toward untamed nature. The English poet William Wordsworth, born in the rugged county of Cumberland, derived "sermons in stone" from his native Lake District. His French contemporary François de Chateaubriand, a self-styled "solitary wanderer," noted that "it is in the prospect of the sublime scenes of nature that the unknown being manifests himself to the human heart."

This romantic love of nature captured the imagination of Europe's wealthy citizens. Trips to the Alps or to The Highlands of Scotland became a popular tonic for those distressed by the brutish scenes of industry around them. More practical steps were taken to curb some of society's most obvious atrocities. The English philosopher Jeremy Bentham voiced a new sensitivity toward animals: "The question is not can they reason, nor can they talk, but can they suffer?" The official response was that at least some of them could. In 1822, the British Parliament passed an act forbidding cruelty to the largest domestic animals (dogs, cats, and birds were not included). Two years later, the Society for the Prevention of Cruelty to Animals became the first such national organization in the world. France formed a similar society in 1845, followed shortly by most other European countries.

Meanwhile, observant members of European colonial administrations in the tropics and elsewhere had begun to voice severe warnings about the effects of rampant imperialism on indigenous flora and fauna. These effects were most apparent in isolated oceanic islands such as Mauritius and Saint Helena, which provided what amounted to laboratory conditions in which to note the depletion of native plants and animals caused by deforestation and overgrazing. As a result, practical measures to combat soil erosion and other forms of despoliation were taken in the Canary Islands, the West Indies, Mauritius, and other islands.

In India, the medical service of the British East India Company compiled a report in 1852 on the "probable effects in an economic and physical point of view of the destruction of tropical forests." The authors pointed to massive soil erosion and the silting up of harbors on the Malabar Coast, and argued that large-scale deforestation could cause a decline in rainfall and, eventually, famine. Serious droughts in the 1830s and later in the 1860s and 1870s appeared to bear them out, and in the wake of the droughts, programs designed to protect the forests were established. Clearly, where it could be proved to the authorities that environmental degradation was not in their economic interests, and where the specters of agricultural failure and social unrest were put before their eyes, they were prepared to listen.

In America, too, when the voice of cool scientific reason was added to Thoreau's passionate pleas, both the government and the general public began to take notice. In 1864, George Perkins Marsh, America's first ambassador to the newly unified kingdom of Italy, published a book entitled *Man and Nature* in which he argued that preservation of the wilderness served both "economical" and "poetical" ends. Forests were not useless, he pointed out. They preserved water resources and prevented soil erosion. Above all, he cautioned against ignorant assaults on the environment: "We can never know how wide a circle of disturbance we produce in the harmonies of nature when we throw the smallest pebble into the ocean of organic life."

The theories of experts in other fields, too, provided ammunition for the preservationists. Most humbling of all were those developed by Charles Darwin in 1859 in his book *On the Origin of Species*, which put an end to the idea that men and women occupied a unique and specially privileged position on earth. Darwin persuasively argued that the human race, for all its superior powers of invention, was not too distant a relative of the great apes.

The German biologist Ernst Haeckel developed Darwin's ideas a stage further. In 1869, he coined the word *oecology* (the *o* was dropped in the twentieth century) to describe the intimate relationship between all living things. Haeckel was proposing that human beings were not merely cousins of the apes, they were working partners of fleas, algae, and rats. In a way, this concept was a modern restatement of the medieval idea of a great chain of being, which also posited links between humans and all other species inhabiting the earth.

Another tenet of traditional knowledge received scientific confirmation at about the same time. "For soil to remain fertile," wrote the German chemist Justus von Liebig in the 1840s, "that which has been taken from it has to be replaced completely." This theory had been blindingly obvious to countless generations of farmers in previous societies, but Liebig now proved it scientifically by analyzing the chemical composition of plants and discovering which substances were removed from the soil when a crop was harvested. Phosphorus, potassium, calcium, magnesium, and nitrogen, he discovered, were the elements most quickly lost when land was cleared without any replenishment. Liebig advocated the use of human manure as a means of revitalizing the soil (and incidentally solving the industrial city's waste-disposal problem), a practice that had been followed for centuries in China and certain other regions of Asia. When this suggestion failed to take hold, he developed mineral nutrients that would allow the same patch of ground to produce crops year after year. The chemical fertilizer industry, which Liebig and others helped inspire, was to become the mainstay of twentieth-century agriculture.

Even those who prized the land only for what they could get out of it now began to understand that they had to give as well as take, and that it was in their interests to protect this precious resource. In 1864, the United States government granted the spectacular Yosemite Valley to the state of California as a public park. Eight years later, in a dramatic concession to the nation's change in mood, almost two million acres of northwest Wyoming became Yellowstone National Park, the first acknowledgment by any government of the need for the large-scale protection of the natural environment. It was a victory for the preservationists, even if a qualified one. It was not until 1894 that hunting was finally prohibited in the park, by which time only 200 of Yellowstone's once-vast number of buffalo remained alive.

The latter provision resulted from the energetic support given to the preservationists by the sportsmen and big-game hunters of America. This was not such an unlikely

The rusting hulks of fishing boats lie beached on salt-encrusted wasteland once covered by the Aral Sea in the Soviet Union's central Asian region. The destruction of the local fishing industry was one result of a failure to assess the environmental impact of drawing off the two major rivers that feed into the sea, the Amu Darya and the Syr Darya, to irrigate land for growing cotton. Between 1960 and 1989, the surface area of the sea diminished by more than 40 percent *(inset map)*. Summer temperatures soared to 115° F., dust storms hazed the sky, and chemical pesticides and defoliants from the cotton fields contaminated supplies of drinking water.

alliance as it might seem. Hunters were often keen and knowledgeable naturalists, and vice versa. During the 1870s, amateur hunters grew increasingly disturbed by the activities of the professionals, whose interests were solely commercial. They drove deer into the water with dogs and clubbed them to death from boats; they baited and trapped wild turkeys and slaughtered birds by the million for their plumage. And now the Plains buffalo, a potent symbol of the frontier, was on the verge of extinction. It was all too much for the sportsmen, whose most outspoken champion was the future president Theodore Roosevelt. The Boone and Crockett Club, which Roosevelt helped found in 1888, pledged "to work for the preservation of the large game of this

country" as well as "to promote manly sport with the rifle." During his two terms in office from 1901 to 1909, Roosevelt oversaw the establishment of five national parks, a professional forest service, and almost 99 million acres of forest reserve.

Not surprisingly, the material value of this land was the chief argument for its protection. Most conservationists had little sympathy with the spiritual concerns of Thoreau or the Romantic poets. A forest should be preserved not because it was beautiful but because it helped the soil retain water and, if properly managed, would serve as a steady source of timber. The attitude of President Roosevelt's chief forester, Gifford Pinchot, was strictly businesslike. "The fundamental principle of the whole conservation policy is that of use," he argued, "to take every part of the land and its resources and put it to use that will serve the most people." And it was this philosophy, rather than that of the idealists, that appealed to the policymakers of the ensuing decades. Pinchot's words were to be precisely echoed in the terms used by the United Nations in 1969 to define conservation as the "rational use of the environment to achieve the highest quality of living for mankind."

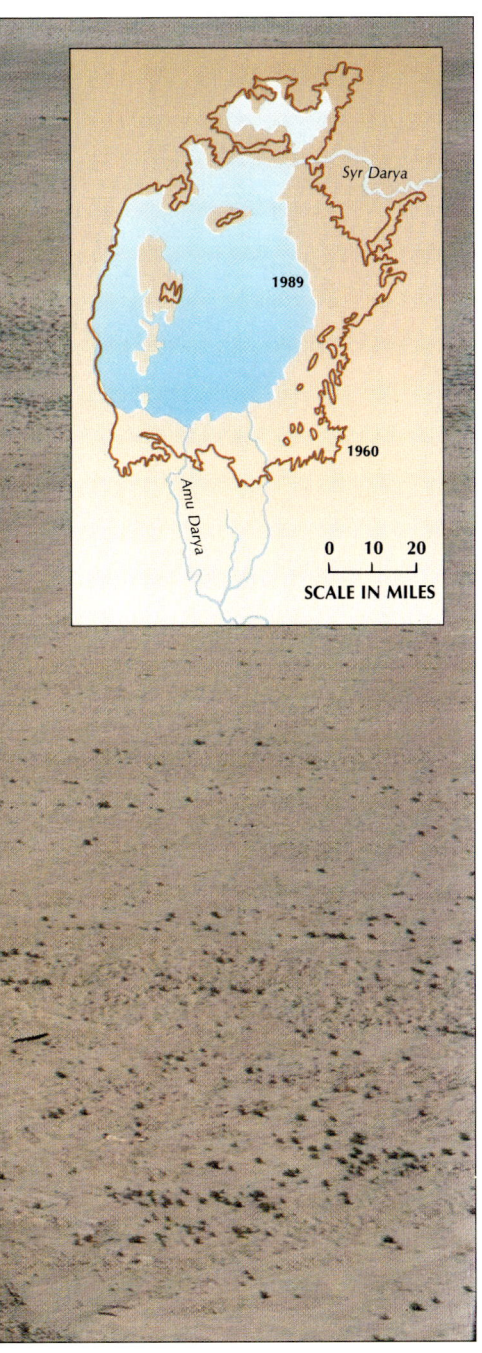

For at least the first half of the twentieth century, the leaders of the world's developed countries were chiefly concerned with either preparing for war or recovering from it, and issues concerning the environment ranked low on their scale of priorities. Isolated gestures were made: Belgium's King Albert I established Albert National Park in the Congo (present-day Zaire) in 1925, giving the gorilla much-needed protection. In the following year, the government of South Africa, with an eye to the profits to be made from tourism, designated almost 8,000 square miles as the Kruger National Park. But a promising international conference for the protection of nature, which assembled in Switzerland in 1913, fell into disarray with the advent of World War I. A similar fate attended attempts at international cooperation during the 1930s—all smashed by the most destructive war the world had ever witnessed.

Meanwhile, the earth's population continued to grow and the wilderness to diminish. Improvements in sanitation and medicine in the developed world resulted in a significant decline in deaths from disease, particularly among children, and between 1850 and 1930, the number of people on the planet doubled from one to two billion. By 1976, scientific advances had extended to the countries of Asia and Africa, and the population had doubled again. To feed these extra mouths, an estimated one billion acres were converted to arable land between 1860 and 1920; another billion acres were put to the plow in the succeeding sixty years. The inevitable overall result was the accelerating depletion of the natural world's living resources.

The application of industrial technology to agriculture increased production and freed many farmers from backbreaking manual labor. Following the introduction of the small gas-driven tractor in the 1920s, for example, the number of hours worked by a modern farmer on a given area of wheat was reduced to one-sixth of the time that his great-grandfather had needed. But the increasing intensity with which land was farmed could have unexpected results, especially when the soil was used for the continuous growing of one type of crop. Golden fields of ripe wheat may give the impression of abundance to hungry human beings, but in reality they are an ecological desert: Very few other species can live in an environment devoted exclusively to the production of a single crop. Even irrigation, intended to bring life to arid places, often has the opposite effect. Where drainage is inadequate, the land can quickly become waterlogged and useless. Salt accumulates rapidly in irrigated regions, either

In April 1985, refugees from the famine-stricken Ethiopian province of Tigre—where three rainless years had left the earth parched and cracked—trek toward hoped-for food and medical relief in the Sudan. In Ethiopia, as elsewhere, famine was a crisis not so much of food shortage and mass starvation as of political mismanagement. While the governments of some of the world's poorest nations, such as Botswana and Costa Rica, managed to provide their citizens with sufficient food, others deliberately exploited inequitable systems of distribution: The Ethiopian regime in the late 1980s bombed 27,500 tons of donated food to prevent it from being eaten by civilians who were not under government control. Most famine deaths occurred from the spread of disease following social breakdown and mass migration, which in Ethiopia included the forced resettlement of people from drought-stricken areas.

from evaporation or from the dissolution of solid salts present in the earth. Because of saturation or salinity, irrigated land is in fact abandoned at about the same rate as it is brought into service.

The environment's response to overexploitation was nowhere more dramatic than in the southern regions of America's Great Plains. In their virgin state, the dry plains of western Kansas, Oklahoma, and northern Texas were covered with a tough, complex carpet of plant life. As many as seventy-five species of short grasses alone might occupy any one area. Even overgrazing by cattle in the late nineteenth century had not destroyed this well-adapted community of vegetation. But the self-styled sodbusters of the early twentieth century were a different matter. There were good profits to be made from the nation's demand for wheat, and so long as prices remained high, the Plains farmers went on busting the sod. By 1925, there were almost two million acres under cultivation in southwestern Kansas; half as much again was dug up during the next five years.

Then disaster struck. From 1931 to 1936, yearly precipitation fell from about nineteen inches to less than twelve. Crops failed over the five worst-affected states. In 1937, almost 250,000 acres were planted with wheat in Cimarron County, Oklahoma; not one was harvested. Lashed by spring winds, the bare, parched soil flew up into the sky. Brown snow fell in New Hampshire; dust settled on ships at sea.

The government responded with the best means it had. Soil conservation experts descended on the Dust Bowl to help those farmers who had not fled westward. Erosion could be discouraged by terracing slopes, the experts advised, or by plowing along the contours of the land rather than in rigid straight lines. Trees should be planted as windbreaks; some land should remain grass-covered. For a while, at least, the chastened farmers paid attention to what the experts had to say, but when the rains resumed and another war broke out in the 1940s, the urge to profit from the land at any cost overwhelmed their better judgment. "What we are doing in the Great Plains today is nothing short of soil murder and financial suicide," admitted the United States secretary of agriculture in 1947, but the murder continued unabated. And when the rains failed, Dust Bowl conditions returned to the region in the mid-1950s, and again in the mid-1970s.

Increasingly, both farmers and conservationists looked to science not only to alleviate ecological damage but also to augment food production. At first, it seemed as if these apparently irreconcilable demands could be met. New strains of high-yield food crops, fed with chemical fertilizers and protected from predators and weeds by a fearsome battery of recently developed pesticides, dramatically increased crop yields in both Western countries and the poorer nations of the Third World. Nuclear power, adapted for commercial purposes after its development for military use during World War II, promised a pollution-free alternative to fossil fuels. But just as satisfied customers across the world were beginning to take the fruits of these developments for granted, the disastrous scale of their destructive side effects became apparent.

Pesticides gave rise to the first cry of alarm. "As crude a weapon as the cave man's club, the chemical barrage has been hurled against the fabric of life," wrote the American biologist Rachel Carson in *Silent Spring*, a passionate attack on the pesticide industry. Writing in 1962, she conjured up the image of a world without songbirds, and indeed birds were falling victim to chlorinated hydrocarbons, such as DDT, which accumulate in the bodies of any organism that consumes them. Birds high up in the food chain contained the highest concentration of pesticides: Their diet

THE HARVEST OF TECHNOLOGY

Scientists examine hybrid strains of rice at the International Rice Research Institute in the Philippines, established with grants from U.S. foundations in 1962. The institute's first major success was a crossbred variety of rice with a yield three times that of traditional plants, which by 1968 was grown in eighty countries.

Seedlings developed by tissue culture—which involves the rearing of cells isolated from living plants—are suspended in test tubes in a sterile growth medium that contains hormones for their regeneration. By using the parts of a plant uncontaminated by disease, tissue culture can produce virus-free crops. It is also a preparatory stage in plant genetic engineering, in which genes are inserted into single cells.

Twentieth-century genetics and biochemistry have enabled the development of fast-maturing plants that allow more crops to be grown each year, as well as the breeding in of genetic resistance to pests and diseases and to harsh conditions such as drought and salinity. In the 1960s, when the effects of this application of science to crop growing became known as the green revolution, many of the newly developed plants required good soil and plentiful amounts of water and fertilizer to produce their increased yields; since then, further advances in biotechnology have increased world food production by 55 billion tons a year. The benefits have been most apparent in parts of Asia and South America, where food supplies have more than kept pace with the population increase; in Africa, progress has so far been slowed by political obstacles and local land tenure systems. But the uses of biotechnology are not always beneficial to Third World countries: The laboratory production of substances, such as cocoa butter, in countries that previously had to import them has the potential to transform world markets—to the likely advantage of the richer nations.

Lettuce plants flourish in a commercial greenhouse without soil, grown by a technique known as hydroponics. The plant roots are immersed either in a water solution containing the minerals necessary for growth, or in gravel or sand to which a nutrient solution is added. This method of cultivation is appropriate for regions that have a favorable climate but are lacking in fertile soil; however, installation costs are high.

of smaller birds, mammals, or fish was already loaded with a deadly chemical cocktail. In the 1960s, fish in Lake Michigan were found to contain 3 to 8 parts per million DDT; in herring gulls, which ate the fish, this figure rose to 3,177. Birds of prey laid eggs with shells so thin they broke in the nest.

While their natural predators died, many species of agricultural pest developed a resistance to the poisons directed against them, forcing scientists to develop yet another chemical weapon, a process that became known as the "pesticide treadmill." One variety of malarial mosquito that appeared in the late 1980s was immune to four of science's most deadly pesticides: DDT, aldrin, dieldrin, and Malathion.

Many countries proceeded to ban these poisons, but other forms of environmental pollution proved more difficult to legislate against. One such problem was the runoff from chemical fertilizers. Adding water-soluble nutrients to the soil increases crop yields, but unfortunately, the crop does not get the full benefit of these chemicals. Excess nitrates and phosphates are washed away into streams, lakes, or ground water. And water enriched by nutrients such as those that come from fertilizers becomes extremely attractive to plant life, particularly to blue-green algae, which has a tendency to multiply at the expense of other aquatic life. A dense growth of algae uses up oxygen at such a rate that fish and other plants are unable to compete. This and other forms of pollution have caused a steep decline in commercial fishing on Europe's inland waterways.

On land, chemical fertilizers can have an even more serious long-term effect. Lakes are surprisingly resilient to pollution by chemicals and often recover their natural balance given sufficient time. The soil may not. A complex mixture of minerals, vegetable matter, and microorganisms, soil is anchored by the roots of plants, which both feed it and feed from it. Dressing a field with manure or vegetable compost both fertilizes the soil and helps to bind together its mineral components. Chemical fertilizers ensure that crops will grow but provide none of the sustaining body. Deprived of its plant cover and its natural humus, the crumbly topsoil begins to blow or wash away. In 1977, it was reported that American farmers were losing almost one inch of topsoil every sixteen years—an amount that might be renewed naturally only every 300 to 1,000 years. As a result, one-third of American agricultural land was judged to be losing productivity.

The dilemmas posed by the use of science and technology to increase food production were not only economic—in terms of balancing short-term gains against long-term losses—but also moral. The latter were especially evident in the case of animals, whose treatment increasingly bore out the law of progress observed by the Irish-born poet Oliver Goldsmith in the eighteenth century: "In all countries, as man is civilized and improved, the lower ranks of animals are oppressed and degraded."

For centuries, those who believed that meat was a natural part of the human diet had scarcely been troubled by ethical concerns. Advocates of vegetarianism had argued that, in the words of one seventeenth-century English writer, it was "unlawful to kill any creature that had life, because it came from God"—but they had attracted few followers. With the introduction of intensive farming methods in the late twentieth century, however, the degradation endured by animals reached a scale that made the moral issues hard to avoid.

Factory farming in Western countries turned poultry, pigs, and to a lesser extent, cattle into utilitarian objects, raw material for the production of meat. On most intensive farms, piglets are removed from their mothers within hours of birth. Without anesthesia, their incisors are clipped and their tails docked, and the males are castrated. Housed in cages with cement or metal-slatted floors, they are kept in darkness and at a constant humid temperature—to induce lethargy and to ensure that food is not used up in keeping them warm. They are injected with a range of antibiotics and have growth-promoting agents added to their feed. They are slaughtered at six to eight months of age. Such treatment may be rationally justified on economic grounds: It yields the maximum amount of meat at the minimum cost. But neither reason nor economics can explain why certain other animal species are treated in an entirely different manner. In the countries of the European Economic Community in the early 1980s, affection was lavished on an estimated 91 million dogs, cats, and other pets. In the United States, families spent $7.5 billion per year on food and veterinary care for an estimated 475 million household pets.

At a personal level, it may be possible to resolve such inconsistencies in cultural attitudes toward nature—by adopting a vegetarian diet, perhaps, or by buying produce exclusively from the small number of farmers who have reverted to traditional methods in growing their crops or raising their animals. At a global level, choices are not made that easily. The natural world is in fact eroded more by human need than by greed, and in the final quarter of the twentieth century, the scale of that need continued to accelerate.

In the fourteen years from 1976 to 1990, the number of people on earth increased by about one billion, bringing the total to slightly more than five billion. Ninety percent of this growth occurred among the developing countries, which had begun to reap the benefits of modern medicine and the expeditious returns of chemical agriculture, while lacking the capacity, and often the desire, to adopt an efficient program of population control. The fastest-growing region of all was Africa: Between 1960 and 1988, the population of that continent more than doubled, from 280 million to more than 620 million. During that same eighteen-year period, Asia's population increased from nearly 1.7 billion to a staggering 3 billion. Although forecasts vary, it is expected that the total population of the earth will double again to around 10 billion by the year 2030.

In China and certain other parts of Asia where population growth actually began to slow, food output more than kept pace with demand. In Africa, on the other hand, where agriculture suffered from a labor shortage because of a drift toward work in the mines and cities, yields were low and variable. And in many countries of the Third World, the pressure on land had tragic consequences. The droughts that struck sub-Saharan Africa in the mid-1980s, killing hundreds of thousands, would have had far less impact had the land not already been overgrazed. In the impoverished and densely populated country of Bangladesh, families migrated onto rich new lands created by silt washed down from the hills. Nothing there protected them against floods from the mountains or murderous inundations from the ocean. In 1970, a tidal wave drowned an estimated 300,000 Bangladeshis on the Ganges-Brahmaputra Delta. Scarcely had the sea receded than thousands more of their compatriots descended to take their place, risking the inevitable onset of future natural disasters.

With population pressure such as this, the preservation of wilderness was not a luxury many countries could seriously contemplate. In the late 1970s, the Brazilian government launched a determined campaign to use its "waste" land for the needs of its expanding population. Rain forests are in fact the world's richest repositories of

In the California desert, 1,818 mirrors, each more than 500 square feet in area, reflect sunlight onto a central receiver. While this experimental field was shut down in the late 1980s, similar projects continued to operate in New Mexico, Spain, and Israel. In the California field, steam from water pumped through the receiver was used to drive a turbine that produced ten megawatts of power for eight hours a day.

life, believed to contain between 50 and 80 percent of all plants and animals, and the rain forests of the Amazon basin constitute the largest remaining system, covering an area equivalent to almost 90 percent of the United States. For the Brazilian government, however, the country's need for agricultural land, and for the hard currency to be derived from exports of beef, overrode international outrage. Cattle ranchers and subsistence farmers advanced along new highways, clearing the land by fire and decimating the native population of Amerindians. In 1990, United Nations experts determined from satellite photographs that as much as 49 million acres of rain forest were being destroyed by fire annually.

Wilderness areas in other parts of the world had their own attendant problems. The East African nation of Kenya has an annual population growth rate of four percent, the highest in the world; yet here, by a twist of irony, economic arguments actually favored the preservation of protected game reserves rather than their conversion to agricultural use. In 1977, an adult male lion in Kenya's Amboseli National Park was worth an estimated $1,150 to a poacher; to the government, its value in tourist revenue was around $515,000. As a result of this persuasive lesson in economics, Kenya banned all sports hunting and attempted to save its endangered elephants by waging a shoot-to-kill war against ivory poachers. But hungry people in need of land for crops may yet do more damage to Africa's wildlife than guns ever have done.

Elsewhere, even those wilderness areas that are too inhospitable for human habitation came under threat. As the world's petroleum and uranium reserves grew scarcer, many believed that limited exploration for natural resources should be permitted in the great virgin landmass of Antarctica. Others argued that humanity is no respecter of limits and opposed any development there whatsoever. They pointed to the case of Alaska, often described as "America's last frontier." When huge oil reserves were discovered at Prudhoe Bay in 1968, oil companies built an 800-mile overland pipeline snaking across the state to the town of Valdez on Alaska's southern coast. In 1989, the tanker *Exxon Valdez* ran aground on rocks in Prince William Sound, blackening about 1,100 miles of Alaska's coastline and destroying the local fishing industry along with an estimated 400,000 seabirds. Six months after the spill and a cleanup operation involving more than 11,000 workers, around 60 percent of the oil still remained on the once-pristine shore.

In the 1980s, it became apparent that even Europe was not immune: The woods and lakes that constitute that continent's own dwindling wilderness were dying from an industrial phenomenon known as acid rain. Sulfur and nitrogen oxides, released from car exhausts or erupting from factory chimneys, dissolve in water to form dilute solutions of sulfuric and nitric acid. These substances descend as rain or snow, often hundreds of miles from the source of emission. In 1982, it was thought that only 8 percent of German forests were affected; two years later, the figure had risen to 50 percent, and the problem had been discovered to exist throughout Europe.

Acid rain brought home to many Europeans the fact that large-scale changes wrought upon the environment by modern technology and the economic policies of the world's richest nations took effect not only in exotic and faraway places but also in their own backyard. For Americans and Eastern as well as Western Europeans, this message was reinforced by leakages of radioactivity from nuclear power stations at Three Mile Island in Pennsylvania in 1979 and at Chernobyl in the Soviet Union in 1986. And the postscript to these latter accidents was that even a change to an alternative form of energy from those upon which Western prosperity had been founded was no guarantee of security.

By the final decade of the century, environmental issues had advanced to the top of the political agenda. The dire predictions of experts gathered at international conferences were increasingly brought to public attention, and governments in the Western world—which were elected by majority vote—were forced to take notice. The word and the color *green* became a media symbol, loaded with political significance. Manufacturers and advertisers, responding to new regulations and with an eye to future market trends, created a demand for products that were claimed to be environmentally safe. When motorists stopped at a garage, they were offered un-

A color-enhanced computer image of the Southern Hemisphere *(opposite)* generated from data collected by a satellite in 1989, shows a hole *(black)* in the stratospheric ozone layer that filters ultraviolet rays from the sun. This radiation increases the incidence of skin cancers and immune deficiencies, and may damage ecosystems. Inert gases that destroy ozone—chiefly CFCs—were banned by some nations in 1978, and timetables for their phasing out were agreed to by other countries in the 1980s. These decisions were prompted by new scientific information that was relayed by pressure groups such as those whose insignias are shown above.

leaded gas, which most understood to have a less-polluting effect than the fuel they had previously bought. In many neighborhoods, containers were provided in which glass bottles and aluminum cans could be deposited for recycling. In supermarkets, certain aerosol spray cans were labeled "ozone friendly." This simplified message, combining science and sentiment, in fact meant that they contained no CFCs, or chlorofluorocarbons, which attack and destroy atmospheric ozone, the gas that prevents excessive amounts of harmful ultraviolet light from reaching the earth. Under the circumstances, however, it was not necessary to understand scientific jargon to know that a crisis was at hand.

Nor did consumers have to study scholarly journals to know what this crisis was about. They could read in the newspapers of new diseases transmitted by the food people ate, food that had been eaten safely for centuries. The greenhouse effect, they were told, which had to do with global warming from the burning of fossil fuels and the prevention of the sun's infrared heat from leaving the atmosphere, was increasing temperatures and would cause a rise in sea levels and the flooding of low-lying areas. Such things would occur, if not in their own lifetimes, then in those of their children or grandchildren. An increasing number of people began to realize that what was at stake was not just the rain forests or the endangered species featured on television documentaries, but many of the things that gave their own lives meaning and that they had so far taken for granted: clean water, fresh food, the parks and open spaces where they took their families on weekends.

As in all times of crisis, it was often difficult to keep a clear sense of perspective. The grand sweep of history appeared to provide little comfort: The decline of Rome and many other empires was caused at least in part by mismanagement and overexploitation of their natural resources. But another factor in their decline was often their succumbing to a mood of inevitable doom. On closer inspection, however, the narrative of human impact on the natural world offers at least as many positive as negative lessons. In the nineteenth century, the North American bison was hunted to the point of extinction; on the other hand, timely human intervention in the form of national parks and conservation laws ensured its survival. In the twentieth century, the use of chemical pesticides caused immense harm to many birds and animals, but strict legislation in Western countries reduced levels of DDT in the environment to almost zero. In the late 1980s, clearance of the Amazonian rain forests increased at an alarming rate—but the bulk of the lowland forests still remains intact, and humans possess the knowledge and the means both to conserve almost all species and to inhibit land degradation and return rainfall to the atmosphere.

Undoubtedly, the economic and political policies of the world's richest nations, while accruing enormous material benefits for their peoples, have caused damage to the natural world. By the 1990s, however, this fact was more widely acknowledged than ever before in the Western world. There was also an increased awareness that, on a global scale, the maintenance of economic and ecological health amounts to the same thing. The problem facing political leaders was how to translate this awareness into action. They could take heart from earlier generations of environmentalists, who had reminded their audiences that humans are not passive victims of deterministic forces but instead have the power to influence their own destiny. "It is never too late to give up our prejudices," advised Thoreau. "What old people say you cannot do, you try and find that you can. Old deeds for old people and new deeds for new."

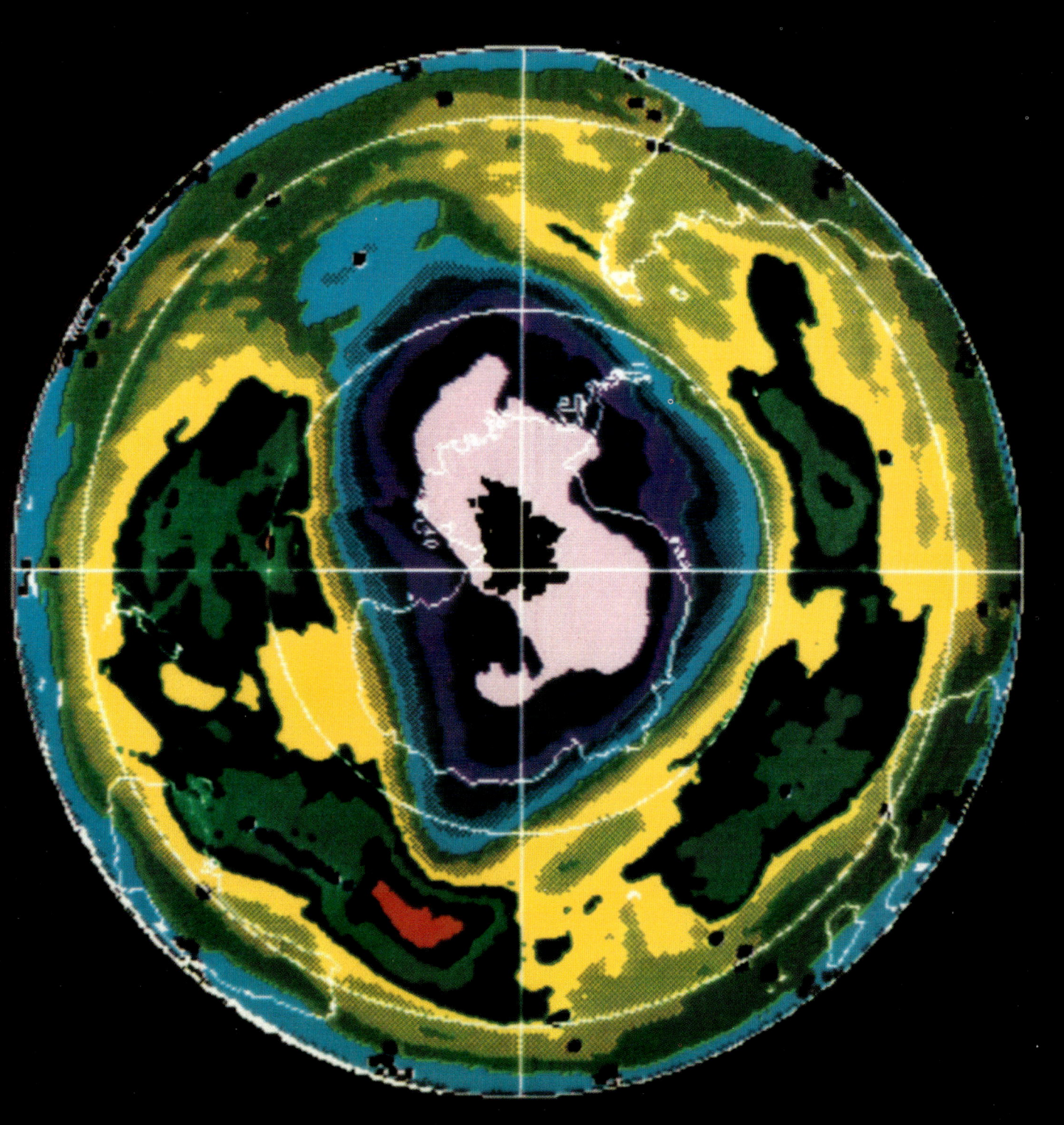

100,000-10,000 BC

Around 100,000 BC the first fully modern humans (Homo sapiens sapiens) emerge, probably in eastern and southern Africa, living by hunting animals and gathering plants.

Bands of human hunter-gatherers move into Eurasia from Africa (40,000 BC), replacing preexisting Neanderthals, and cross the sea gap from Southeast Asia into New Guinea and Australia.

From 20,000 BC, the use of new weapons—bows and arrows, spear-throwers—and tactics such as ambushes and the use of fire to stampede herds increase the hunting efficiency of early humans.

Paintings of animals are executed in caves in southwestern France and northern Spain (15,000 BC).

Grindstones are used to make flour from wild grass seeds in Upper Egypt (12,000 BC).

Hunter-gatherer bands reach the tip of South America (10,000 BC).

Climatic change and the scale of human hunting contribute to the extinction of species such as the mammoth, woolly rhinoceros, and the American horse.

10 million

10,000 BC-0

The cultivation of domesticated wheat, barley, and beans and the herding of livestock (cattle, sheep, goats, pigs) spread outward from the Fertile Crescent (8000 BC). Agriculture develops independently in Southeast Asia and the Americas.

Widespread sea and land trade in precious stones and shells develops (7000 BC).

North African fishing and cattle cultures flourish before the drying-out of the Sahara (4000 BC).

Cities, founded on agricultural prosperity, are built in Mesopotamia (3500 BC). Shortly afterward, urban centers are founded in Egypt.

Trade and agriculture bring wealth to the palaces of Minoan Crete and Mycenaean Greece (2000 BC).

Horses are domesticated; pastoral nomads range across the Asian steppes with flocks of sheep and goats (1500 BC).

Cities supported by rice-growing are established in the Ganges Valley in India (800 BC).

Greek and Phoenician colonies are established around the Mediterranean (750 BC); Athens and other city-states flourish on the Greek mainland (fifth century BC).

The Roman republic is established (550 BC).

The Chinese empire is unified under the Qin dynasty (221 BC).

Rome becomes dependent on the import of grain from Sicily, Sardinia, and after the destruction of Carthage in 146 BC, North Africa.

200 million

0-AD 1000

The growing population of Rome, reaching one million in the first century AD, spurs further military conquests. By the second century, the Roman Empire extends from Spain and Britain in the West to Syria in the East.

Tropical forest civilizations develop in lowland Mexico and Southeast Asia.

The Bantu peoples spread across sub-Saharan Africa.

Paper is manufactured in China (second century).

Rome is sacked by the Visigoths (410).

Mesopotamian irrigation systems help support a growing population (500-800).

Arab armies spread the Islamic religion throughout the Middle East and North Africa (seventh century); through the medium of Arab scholars, the classical learning of Greece and Rome is later passed on to Europe.

In China, the Grand Canal is constructed between the Yangtze and Yellow rivers to transport grain from the south to the north (seventh century).

Viking seafarers settle in Iceland (870) and Greenland.

The Polynesians, having colonized most oceanic islands in the Pacific, settle in New Zealand (ninth century).

300 million

TimeFrame 100,000 BC-AD 1990

AD 1000-1500

In Europe, favorable climatic conditions from the eleventh century promote advances in agriculture and an increase in population. Crop rotation and open fields divided into strips for efficient plowing are introduced. Wine from the south is traded for grain from the north.

In China, the introduction of a fast-maturing rice from Vietnam in the eleventh century allows more than one harvest per year. China also benefits from improved technology such as locks in canals, machines for textile production, and the invention of movable-type printing.

Windmills are in use in Europe from 1180.

In the highlands of Central and South America, Aztec and Inca civilizations reach their apogee (1300-1500).

The Black Death, beginning in Asia, reaches Europe in 1347 and kills more than one-third of the Continent's population.

African sub-Saharan civilizations flourish in Zimbabwe (fourteenth century) and West Africa (Ghana, Songhai).

Open fields in Europe are enclosed for sheep farming from the fifteenth century.

Johannes Gutenberg produces the first printed books in Europe (1455).

The voyages of European navigators prepare the way for the global expansion of European power: Bartolomeu Dias sails around the Cape of Good Hope (1488); Christopher Columbus crosses the Atlantic and makes landfall in the Caribbean (1492); Vasco da Gama reaches India (1498).

400 million

AD 1500-1990

The transportation of slaves from West Africa to the Americas begins (1505).

Spanish conquistadors establish a colonial empire in Central and South America (1519-1535); European diseases kill off 90 percent of the indigenous populations.

The potato is imported to Europe from South America (1525). Other imports from the New World include corn, tobacco, and the turkey. Coffee and tea are introduced to Europe from the Orient.

The first permanent English settlement in North America is established (1607), followed shortly by French and Dutch settlements.

The Little Ice Age, beginning around 1590, causes failed harvests and abandonment of uplands in Europe.

The introduction of seed drills and similar technology in the early eighteenth century signals the beginning of the mechanization of European agriculture.

Steam engines are developed in Britain for use in coal mines (1765).

The establishment of a penal colony at Sydney Cove in 1788 inaugurates the British settlement of Australia.

Steam-powered trains are introduced in Britain (1825); extensive railroad networks are built in Europe and America, and subsequently in Australia, Africa, India, Japan, and Russia.

The Industrial Revolution in Europe and later in America promotes the growth of towns and accelerates the impact of human technology on the natural world. From the 1830s, the mechanization of Western agriculture is speeded by the introduction of the mechanical reaper, the steam plow, and the combine harvester.

The first oil well begins operation in Pennsylvania (1859).

Following advocacy for conservation, Yosemite Valley in California becomes the first national park (1864).

Overcultivation of the American plains leads to drought and the creation of Dust Bowl conditions in the 1930s.

The use of chemical pesticides becomes widespread in Western agriculture in the 1940s and 1950s; strict legislation is enforced from the 1960s to limit their harmful side effects.

In China, the Communist government from 1949 institutes radical programs of collectivization and population control.

The first nuclear power station, offering an alternative energy source to fossil fuels, is opened in the United Kingdom (1956).

Genetic engineering is increasingly applied to crop growing and stockbreeding. The International Rice Research Institute is founded in the Philippines (1962) to develop new strains of rice with increased yields.

The consequences of environmental deterioration in the 1980s—including famine in Africa, global warming, and the depletion of the earth's ozone layer—increase awareness in the Western world of the need for long-term management of natural resources.

5 billion +

The chronology on the left lists some of the main events in the narrative of human impact on the natural world. The figures at the foot of each column represent the approximate human population of the world at the end of the given timespan.

Population growth has always impressed itself on the landscape—by prompting migrations to previously uninhabited regions, by causing unused land to be put to the plow, or by spurring the development of new technologies. In the nineteenth and twentieth centuries, however, the curve rose more and more steeply: The population increase from 1950 to 1990 alone equaled the total increase from the emergence of the human species to 1950. Although the rate of increase began gradually to slow, the absolute population figure continued to rise rapidly. The 1990 population level was at once a measure of human success in deriving a living from the planet and a challenge to the flexibility of political and economic structures.

PICTURE CREDITS

The sources for the illustrations that appear in this book are listed below. Credits from left to right are separated by semicolons; from top to bottom they are separated by dashes.

Cover: Photograph of oilseed rape fields in Fife, Scotland, Patricia Macdonald, Midlothian. **2-3:** Robert Harding Picture Library, London. **8:** James Wellard / Sonia Halliday Photographs, Weston Turville, Buckinghamshire. **10, 11:** Jane Taylor / Sonia Halliday Photographs, Weston Turville, Buckinghamshire; Trustees of the British Museum, London. **13:** Archivo Oronoz, Madrid. **14:** Trustees of the British Museum, London. **15:** Robert Harding Picture Library, London. **16:** Werner Forman Archive, London / Statens Historiska Museum, Stockholm; Horniman Museum, London. **18-19:** Art by Alan Hollingbery. **20, 21:** V. Southwell / Robert Harding Picture Library, London; Hirmer Verlag, Munich. **22:** Hirmer Verlag, Munich (background); Pelizaeus Museum, Hildesheim. **24:** Trustees of the British Museum, London. **25:** Werner Forman Archive, London / Schimmel Collection, New York City. **26:** Map by Alan Hollingbery. **27:** G. Dagli Orti, Paris. **28, 29:** Roland and Sabrina Michaud / John Hillelson Agency, London; Erich Lessing, Vienna / Trustees of the British Museum, London. **31:** Trevor Page / The Hutchison Library, London. **32-33:** David Burnett / Colorific, London. **34-35:** Marion Morrison / South American Pictures, Woodbridge, Suffolk, England. **36-37:** © John Glover 1979, Godalming, Surrey. **38:** G. Dagli Orti, Paris /courtesy Musée du Louvre, Paris. **40:** Elliott Erwitt / Magnum, London; Trustees of the British Museum, London. **42:** Scala, Florence / Museo Archeologico, Venice. **43:** Trustees of the British Museum, London. **44:** Erich Lessing, Vienna / National Museum of Damascus, Syria; Trustees of the British Museum, London. **45:** Trustees of the British Museum, London. **46:** Brooklyn Museum, New York. **49:** © Michael Holford, Loughton, Essex / Collection British Museum, London. **50:** Vatican Library, Rome—Trustees of the British Museum, London (3). **52:** Roger Wood, Deal, Kent, England. **54:** Trustees of the British Museum, London; Vatican Library, Rome. **55:** Trustees of the British Museum, London. **56:** Osterreichische Nationalbibliothek, Vienna. **58:** Collection of Tokyo National Museum. **60:** Paul Dix / Susan Griggs Agency, London. **61:** Courtesy the Board and Trustees of the Victoria and Albert Museum, London. **62-63:** Réunion des Musées Nationaux, Paris. **65:** E. T. Archive, London. **66-67:** By permission of the British Library, London, OR 2362 (4). **69:** By permission of the India Office Library (British Library), London, NHD 43 f.76. **70, 71:** © Wan-go H. C. Weng, Lyme, New Hampshire / Collection Palace Museum, Beijing; Rene Burri / Magnum, London. **73:** © Brian J. Coates / Bruce Coleman Limited, London. **74:** Museum of Fine Arts, Boston. **75:** Vatican Library, Rome. **77:** © Takeji Iwamiya / Pacific Press Service, Tokyo. **79:** Salgado / Magnum, London. **80:** Staatliche Museen zu Berlin; L. A. East, Liphook, Hampshire—Trustees of the British Museum, London. **81:** Staatsgalerie, Stuttgart. **82, 83:** Eileen Tweedy / courtesy the Science Museum, London (2)—Ann Ronan Picture Library, Taunton, Somerset (2); Board of Trustees of the National Museums and Galleries on Merseyside (Walker Art Gallery). **84, 85:** Drake Museum, courtesy the American Petroleum Institute, Washington, D.C.—British Nuclear Fuels plc, Warrington, Cheshire; Leif Berge, Hinna, Norway. **86:** Musée Des Arts Décoratifs, Paris. **88, 89:** Buckinghamshire Record Office, Aylesbury; by permission of the British Library, London, Roy 2B VII, ff112 and 155v. **91:** Bodleian Library, Oxford, Bodley 264, f81v. **92, 93:** Scala, Florence / Biblioteca Marciana, Venice (4). **94:** Erich Lessing, Vienna / Graphische Sammlung Albertina, Vienna. **95:** Bibliothèque Nationale, Paris. **97:** By permission of the British Library, London, Harl. 1585. **98:** Images Colour Library, Leeds; Bridgeman Art Library, London, courtesy the British Museum, London. **99:** Bridgeman Art Library, London, courtesy the Kunsthistorisches Museum, Vienna; Hubert Josse, Paris. **100:** Fitzwilliam Museum, Cambridge, England. **103:** By permission of the British Library, London, Harl. 4425, 12v. **105:** Courtesy the Board and Trustees of the Victoria and Albert Museum, London. **106, 107:** Trustees of the British Museum, London (2); Varga / Artephot, Paris; Antivarisk Topografiska Arkivet, Stockholm—National Palace Museum, Beijing. **108-109:** Scala, Florence / Ca'Rezzonico, Venice. **110:** Rijksdienst Beeldende Kunst on loan to the Frans Halsmuseum de Hallen, Haarlem. **112, 113:** Library of Congress, Washington, D.C.; © Tom Bean / DRK Photo, Sedona, Arizona. **115:** Print Collection Miriam and Ira D. Wallach, Division of Art, Print and Photographs, New York Public Library, Astor, Lenox and Tilden Foundations. **116:** Trustees of the British Museum (Natural History), London. **117:** E. T. Archive, London; Eileen Tweedy /by permission of the Linnean Society of London—Trustees of the British Museum (Natural History), London. **118:** Trustees of the British Museum (Natural History), London; E. T. Archive, London / Trustees of the British Museum, London. **119:** Trustees of the British Museum (Natural History), London (3). **120, 121:** Robert Harding Picture Library, London / Collection National Galerie, Berlin; Archiv für Kunst und Geschichte, Berlin; Trustees of the British Museum (Natural History), London. **122:** National Gallery, London. **123:** Palace Museum, Beijing; Manchester City Art Galleries. **124, 125:** British Architectural Library, RIBA, London (2). **126:** Derby Museums and Art Gallery. **129, 130, 131:** Bridgeman Art Library, London / Collection Cheltenham Art Gallery and Museum; Tate Gallery, London. **133:** By permission of the British Library, London, 1038.b.24. **134:** By permission of the British Library, London, 45.i.3.4, p873; The Mansell Collection, London—Photothèque du Musée de l'Homme, Paris; Mary Evans Picture Library, London. **135:** Scala, Florence / Ca'Rezzonico, Venice; D. Graf / Staatliche Museen Preussischer Kulturbesitz Museum für Völkerkunde, Berlin—Trustees of the British Museum (Natural History), London—© 1985 The Metropolitan Museum of Art, New York City. **136:** Sonia Halliday Photographs, Weston Turville, Buckinghamshire; courtesy the Board and Trustees of the Victoria and Albert Museum, London—Bodleian Library, Oxford, John Johnson Collection, De Bry folder I—Inventutis exercitia, pl. XXXVI. **137:** Peter Newark's American Pictures, Bath—Royal Botanic Gardens, Kew, London; by permission of the India Office Library (British Library), London, NHD 19, f8. **138:** Courtesy the Board and Trustees of the Victoria and Albert Museum, London; by permission of the India Office Library (British Library), London—by permission of the British Library, London, 447.g.4, p252; Mary Evans Picture Library, London. **139:** Wellcome Institute Library, London—Royal Botanic Gardens, Kew, London—National Library of Medicine, Bethesda, Maryland; by permission of the British Library, London, 451.i.3.4, p419. **140:** By Ralph Perry, © National Geographic Society, Washington, D.C. **142:** By permission of the British Library, London, 713.i.1.11 F / P. **143:** Museum Folkwang, Essen. **144:** Trustees of the British Museum (Natural History), London. **145:** Courtesy the Kendall Whaling Museum, Sharon, Massachusetts. **146:** Foreign and Commonwealth Office, London. **147:** © 1990 Monas Hierogliphica, Milan. **148, 149:** By Steve McCurry, © National Geographic Society, Washington, D.C.; Steve McCurry / Magnum, London. **150, 151:** Archivio di Stato, Venice (3); National Museum of American Art, lent by the U.S. Department of the Interior / Art Resource, New York City; G. Ziesler / Bruce Coleman Limited, London. **154, 155:** By David Turnley, © National Geographic Society, Washington, D.C.; map by Town Group Consultancy. **156:** Sebastiao Salgado / Magnum, London. **158:** Steve McCurry / Magnum, London— Charlotte Raymond / Science Photo Library, London. **159:** Hank Morgan / Science Photo Library, London. **162-163:** Peter Menzel / Science Photo Library, London. **164:** Greenpeace, London—Friends of the Earth, London—World Wild Fund for Nature, Godalming, Surrey—Greenpeace, London. **165:** NASA, Goddard Space Flight Center, Greenbelt, Maryland.

BIBLIOGRAPHY

Anderson, Perry, *Passages from Antiquity to Feudalism.* London: New Left Books, 1974.
Anglicus, Bartholomew, *Medieval Lore.* London: Elliot Stock, 1893.
Arnold, T. W., *The Legacy of Islam.* London: Oxford University Press, 1965.
Attenborough, David, *The First Eden: The Mediterranean World and Man.* London: Collins/BBC Books, 1987.
Austin, M. M., and P. Vidal-Naquet, *Economic and Social History of Ancient Greece: An Introduction.* London: B. T. Batsford, 1977.
Aykroyd, Wallace R., *Sweet Malefactor: Sugar, Slavery and Human Society.* London: Heinemann, 1967.
Barker, Graeme, *Prehistoric Farming in Europe.* Cambridge: Cambridge University Press, 1985.
Bazarov, Konstantin, *Landscape Painting.* London: Octopus, 1981.
Berrall, Julia S., *The Garden: An Illustrated History.* London: Penguin, 1978.
Bloch, Marc, *Feudal Society.* London: Routledge & Kegan Paul, 1989.
Blum, Jerome, ed., *Our Forgotten Past: Seven Centuries of Life on the Land.* London: Thames and Hudson, 1982.
Blunt, Wilfrid, *The Compleat Naturalist: A Life of Linnaeus.* London: Collins, 1984.
Boardman, John, Jasper Griffin and Oswyn Murray, eds.:
  The Oxford History of the Classical World: Greece and the Hellenistic World. New York: Oxford University Press, 1988.
  The Oxford History of the Classical World: The Roman World. New York: Oxford University Press, 1988.
Braudel, Fernand, *Capitalism and Material Life, 1400-1800.* Transl. by Miriam Kochan. New York: Harper & Row, 1974.
Bray, Francesca, "Agriculture." In *Science and Civilisation in China,* ed. by Joseph Needham, Vol. 6. Cambridge: Cambridge University Press, 1984.
Brothwell, D., and P. Brothwell, *Food in Antiquity.* New York: Praeger, 1969.

Browning, Robert, ed., *The Greek World.* London: Thames and Hudson, 1985.
Buchanan, Keith, *The Transformation of the Chinese Earth.* New York: Praeger, 1970.
Carson, Rachel, *Silent Spring.* Boston: Houghton Mifflin, 1987.
Coates, Austin, *The Commerce in Rubber: The First 250 Years.* New York: Oxford University Press, 1987.
Cotterell, Yong Yap, and Arthur Cotterell, *The Early Civilization of China.* London: Weidenfeld and Nicolson, 1975.
Cressey, George B., *Land of the 500 Million: A Geography of China.* New York: McGraw-Hill, 1955.
Crosby, Alfred W., *Ecological Imperialism: The Biological Expansion of Europe, 900-1900.* New York: Cambridge University Press, 1986.
Cummins, John, *The Hound and the Hawk: The Art of Medieval Hunting.* London: Weidenfeld and Nicolson, 1988.
Cunliffe, Barry, *Rome and Her Empire.* London: Bodley Head, 1978.
Delano-Smith, Catherine, *Western Mediterranean Europe: A Historical Geography of Italy, Spain and Southern France Since the Neolithic.* London: Academic Press, 1979.
Dilke, O. A. W., *Greek and Roman Maps.* London: Thames and Hudson, 1985.
Douglas, Mary, *Purity and Danger: An Analysis of the Concepts of Pollution and Taboo.* New York: Ark Paperbacks, 1984.
Duby, Georges, *Rural Economy and Country Life in the Medieval West.* Columbia, South Carolina: University of South Carolina Press, 1990.
Edgerton, Samuel, Jr., *The Renaissance Discovery of Linear Perspective.* New York: Harper & Row, 1978.
Edlin, H. L., *Man and Plants.* London: Aldus, 1967.

Ehrlich, Anne H., and Paul R. Ehrlich:
  Earth. London: Thames Methuen, 1987.
  Extinction: The Causes and Consequences of the Disappearance of Species. New York: Random House, 1981.
Elvin, Mark, *The Pattern of the Chinese Past.* Stanford, California: Stanford University Press, 1973.
The Epic of Gilgamesh. Transl. by N. K. Sandars. London: Penguin, 1972.
Fagan, Brian M., *People of the Earth: An Introduction to World Prehistory.* Glenview, Illinois: Scott Foresman, 1989.
Farb, Peter, *Humankind.* London: Triad Paladin, 1978.
Galloway, J. H., *The Sugar Cane Industry: An Historical Geography from Its Origins to 1914.* New York: Cambridge University Press, 1989.
Gamble, Clive, *The Palaeolithic Settlement of Europe.* New York: Cambridge University Press, 1986.
Garnsey, Peter, *Famine and Food Supply in the Graeco-Roman World: Responses to Risk and Crisis.* New York: Cambridge University Press, 1988.
Gimpel, Jean, *The Medieval Machine: The Industrial Revolution of the Middle Ages.* London: Victor Gollancz, 1977.
Glaber, Radulfus, *The Five Books of Histories.* Ed. and transl. by John France. New York: Oxford University Press, 1990.
Glacken, Clarence, *Traces on the Rhodian Shore: Nature and Culture in Western Thought from Ancient Times to the End of the Eighteenth Century.* Berkeley: University of California Press, 1967.
Goudie, Andrew, *The Human Impact on the Natural Environment.* Oxford: Basil Blackwell, 1986.
Gowlett, John, *Ascent to Civilization.* London: Collins, 1984.
Harris, Marvin, *Cows, Pigs, Wars and Witches: The Riddles of Culture.* New York: Vintage, 1978.
Hartley, Dorothy, *Lost Country Life: How English Country Folk Lived, Worked, Threshed, Thrashed.* New York: Pantheon, 1981.

Hawkes, David, transl., *The Songs of the South: An Ancient Chinese Anthology of Poems.* New York: Penguin, 1985.
Healy, John F., *Mining and Metallurgy in the Greek and Roman World.* London: Thames and Hudson, 1978.
Hesiod, *Theogony* and *Works and Days.* Transl. by Dorothea Wender. London: Penguin, 1973.
Honour, Hugh, *The New Golden Land: European Images of America from the Discoveries to the Present Time.* London: Allen Lane, 1976.
Hook, Brian G., ed., *The Cambridge Encyclopedia of China.* New York: Cambridge University Press, 1982.
Hunter, J. M., *Land into Landscape.* London: George Godwin, 1985.
Jenkins, Ian, *Greek and Roman Life.* London: British Museum Publications, 1986.
Jones, E. L., *The European Miracle: Environments, Economies, and Geopolitics in the History of Europe and Asia.* New York: Cambridge University Press, 1987.
Khazanov, A. M., *Nomads and the Outside World.* Transl. by Julia Crookenden. New York: Cambridge University Press, 1984.
Klingender, Francis, *Animals in Art and Thought to the End of the Middle Ages.* Ed. by Evelyn Antal and John Harthan. London: Routledge & Kegan Paul, 1971.
Komarov, Boris, *The Destruction of Nature in the Soviet Union.* White Plains, New York: Sharpe, 1980.
Kuhn, Dieter, "Textile Technology: Spinning and Reeling." In *Science and Civilisation in China,* ed. by Joseph Needham, Vol. 5. Cambridge: Cambridge University Press, 1988.
Ladurie, Emanuel Le Roy, *Montaillou: Cathars and Catholics in a French Village, 1294-1324.* Transl. by Barbara Bray. London: Penguin, 1984.
Lawson-Peebles, Robert, *Landscape and Written Expression in Revolutionary America: The World Turned Upside Down.* New York: Cambridge University Press, 1988.
Lee, R. B., and I. DeVore, eds., *Kalahari Hunter-Gatherers: Studies of the !Kung San and Their Neighbors.* Cambridge, Massachusetts: Harvard University Press, 1976.

Lehner, Ernst, and Johanna Lehner, *Folklore & Odysseys of Food & Medicinal Plants.* London: Harrap, 1973.
Leroi-Gourhan, A., *The Art of Prehistoric Man in Western Europe.* London: Thames and Hudson, 1968.
Lewington, Anna, *Plants for People.* London: Natural History Museum Publications, 1990.
Loewe, Michael, *The Pride That Was China.* New York: St. Martin's Press, 1990.
Longone, Janice B., *Mother Maize and King Corn: The Persistence of Corn in the American Ethos.* Ann Arbor: The William L. Clements Library, University of Michigan, 1986.
Mackay, Angus, *Spain in the Middle Ages: From Frontier to Empire, 1000-1500.* London: Macmillan, 1977.
McKeown, Thomas, *The Modern Rise of Population.* London: Edward Arnold, 1976.
McLuhan, T. C., *Touch the Earth: A Self-Portrait of Indian Existence.* New York: Simon & Schuster, 1971.
McNeill, William, *Plagues and Peoples.* London: Penguin, 1979.
Meadows, Jack, ed., *The History of Scientific Discovery.* Oxford: Phaidon, 1987.
Meiggs, Russell, *Trees and Timber in the Ancient Mediterranean World.* Oxford: Clarendon Press, 1982.
Mennell, Stephen, *All Manners of Food: Eating and Taste in England and France from the Middle Ages to the Present.* Oxford: Basil Blackwell, 1985.
Nash, Roderick, *Wilderness and the American Mind.* New Haven: Yale University Press, 1982.
Pepper, David, *The Roots of Modern Environmentalism.* London: Croom Helm, 1984.
Perlin, John, *A Forest Journey: The Role of Wood in the Development of Civilization.* New York: W. W. Norton, 1989.
Pfeiffer, John E., *The Emergence of Humankind.* New York: Harper & Row, 1985.

Postgate, Nicholas, *The First Empires.* Oxford: Elsevier / Phaidon, 1977.
Pounds, Norman J. G.:
Hearth & Home: A History of Material Culture. Bloomington and Indianapolis: Indiana University Press, 1989.
An Historical Geography of Europe 450 BC to AD 1330. Cambridge: Cambridge University Press, 1973.
Pullan, Brian, ed., *Crisis and Change in the Venetian Economy in the Sixteenth and Seventeenth Centuries.* London: Methuen, 1968.
Reader, John, *Man on Earth.* London: Penguin, 1990.
Reischauer, E. O., ed., *Ennin's Travels in Tang China.* New York: Ronald Press, 1955.
Rickman, Geoffrey E., *The Corn Supply of Ancient Rome.* Oxford: Clarendon Press, 1980.
Roberts, Neil, *The Holocene: An Environmental History.* Oxford: Basil Blackwell, 1989.
Russell, W. M. S., *Man, Nature and History.* London: Aldus, 1967.
Salaman, Redcliffe N., *The History and Social Influence of the Potato.* Cambridge: Cambridge University Press, 1949.
Sauer, Carl Ortwin, *Land and Life.* Ed. by John Leighly. Berkeley and Los Angeles: University of California Press, 1963.
Seymour, John, and Herbert Girardet, *Far from Paradise: The Story of Human Impact on the Environment.* London: Green Print, 1990.
Simmons, I. G., *Changing the Face of the Earth: Culture, Environment, History.* Oxford: Basil Blackwell, 1989.
Smith, Bernard, *European Vision and the South Pacific.* New Haven: Yale University Press, 1988.
Spence, Jonathan D., *The Death of Woman Wang.* New York: Penguin, 1979.
Steven, Margaret, *First Impressions: The British Discovery of Australia.* London: British Museum (Natural History), 1988.
Stockwell, Christine, *Nature's Pharmacy: A History of Plants and Healing.* London: Arrow, 1989.
Tannahill, Reay, *Food in History.* London: Eyre Methuen, 1973.

Thacker, Christopher, *The History of Gardens.* London: Croom Helm, 1979.
Thesiger, Wilfred, *Arabian Sands.* London: Penguin, 1964.
Thomas, Keith, *Man and the Natural World: Changing Attitudes in England, 1500-1800.* London: Penguin, 1984.
Thoreau, Henry D., *Walden.* Ed. by J. Lyndon Shanley. Princeton: Princeton University Press, 1971.
Tyler, Ron, *Visions of America: Pioneer Artists in a New Land.* London: Thames and Hudson, 1983.
Vainshtein, Sevian, *Nomads of Southern Siberia: The Pastoral Economies of Tuva.* Ed. by Caroline Humphrey, transl. by Michael Colenso. New York: Cambridge University Press, 1983.
van Andel, Tjeerd H., and Curtis Runnels, *Beyond the Acropolis: A Rural Greek Past.* Stanford, California: Stanford University Press, 1987.
van der Post, Laurens, *The Lost World of the Kalahari.* London: Penguin, 1962.
Walgate, Robert, *Miracle or Menace? Biotechnology and the Third World.* London: The Panos Institute, 1990.
Waters, Frank, *The Book of the Hopi.* New York: Penguin, 1977.
Webster, C. C., and W. J. Baulkwill, *Rubber.* Harlow, Essex: Longman, 1989.
White, K. D., *Greek and Roman Technology.* London: Thames and Hudson, 1984.
White, Lynn, *Medieval Technology and Social Change.* New York: Oxford University Press, 1962.
Wolman, M. G., and F. G. A. Fournier, eds., *Land Transformation in Agriculture.* New York: Wiley, 1988.
Worster, Donald:
Dust Bowl: The Southern Plains in the 1930s. New York: Oxford University Press, 1982.
Nature's Economy: A History of Ecological Ideas. New York: Cambridge University Press, 1985.
Worster, Donald, ed., *The Ends of the Earth: Perspectives on Modern Environmental History.* New York: Cambridge University Press, 1989.

ACKNOWLEDGMENTS

The following materials have been reprinted with the kind permission of the publishers: Page 9: "Here every man . . ." quoted from *Arabian Sands* by Wilfred Thesiger, London: Penguin, 1964. Page 17: "For about two minutes . . ." quoted from *The Lost World of the Kalahari* by Laurens van der Post, London: Penguin, 1962. Page 62: "The people live . . ." quoted from *The Songs of the South: An Ancient Chinese Anthology of Poems* transl. by David Hawkes, London: Penguin, 1985. Page 68: "Ten tubes upon each wheel . . ." quoted in *The Pattern of the Chinese Past* by Mark Elvin, Stanford, California: Stanford University Press, 1973. Page 75: "destitute and ravaged . . ." and "On top of all this . . ." quoted in *The Death of Woman Wang* by Jonathan D. Spence, New York: Penguin, 1979. Page 87: "After men had eaten . . ." and "At the millennium . . ." quoted from *The Five Books of Histories* by Radulfus Glaber, ed. and transl. by John France, New York: Oxford University Press, 1990.

The editors also wish to thank the following individuals and institutions for their valuable assistance in the preparation of this volume:
England: London—Michael Donaghy; Anne-Marie Ehrlich; Rupert Hastings, The Royal Botanic Gardens, Kew; Beth McKillop, Oriental Collections, British Library; Christine Noble; Andrew Norris, RIBA Drawings Collection; Chris Rawlings, Photographic Services, British Library. Oxford—Suzanne Williams.
United States: Sharon, Massachusetts—Stuart M. Frank, The Kendall Whaling Museum.

INDEX

Numerals in italics indicate an illustration of the subject mentioned.

A

Aborigines, 32, 46, 128-129
Achilles (Greek hero), 47
Acid rain, 141, 163
Aegean Sea, 47
Afghanistan, 28
Africa, 12, 16; colonization of, 122-125; crops in, 120, 146; food supply in, 158; gold mining in, 79; hunter-gatherers in, 17-19; landscape in, 125, 147; life expectancy in, 59; nomads in, 25; population growth in, 151, 163; railroads in, 146; slaves in, 116, 120, 124. See also North Africa; South Africa
Africans, in North America, 132
Agricola, Georgius (German mineralogist), 82
Agriculturalists, 11, *map* 26; Iroquois as, 22; in Mexico, 21-23; rise of, 19-21; settlement of, 27; and wheat, *20-21*
Agriculture: in Africa, 161; in Central America, 21-23; chemical fertilizers in, 153; in China, 24-25, 59-60, 61, *62-63,* 66-78; in Corfu, *40;* in Crete, 42-43; in Egypt, *136;* in England, 89, *129;* in Europe, 87-89, *91, 92-93,* 94-99, 104, *135;* in Fertile Crescent, 41; in Greece, *38, 39,* 46-49; inventions in, 128; in Iran, 29; in Iraq, *20-21;* in Italy, 57; in Japan, *58;* in Mexico, 113-114; in Middle Ages, 90; in Mycenae, 44-45; in New World, *135;* in North America, *135;* in Peru, *134;* and pesticides, 157-160; in Roman Empire, 53-54; in Rome, 49-51; in Sicily, *134;* in Slavic lands, 95; in Sumer, 42; in Third World, 145-146; in United States, 141; in Vietnam, 72
Agrimony plant, *56*
Ainu (people), 46
Akbar (emperor of India), *105*

Alaska, 163
Albert I (king of Belgium), 155
Albert National Park (Congo), 155
Albertus Magnus (German theologian), 102
Algeria, 9; mosaic from, *52*
Alps (mountains), 89, 94-95, 143, 152
Altamira (Spain), cave paintings at, 17
Amboseli National Park (Kenya), 163
Amerindians, 131, 162; crops of, 119-120; lifestyle of, 121, 132, 149-152; and smallpox, 118-119; and tobacco, 138; and venereal diseases, 120
Andes (mountains), 21, 139
Andromeda, *117*
Anglicus, Bartholomaeus (Franciscan friar), quoted, 101-102
Angola, banana plantation in, *147*
Animals: in Africa, 17-18, 127; in Australia, *116, 118;* in China, 63, 71; domestication of, 19-21, 27, 114; in England, *131;* in Europe, 13, 94; extinction of, *52, 54;* and factory farming, 160-161; in farming, 95; hunting of, *105-107;* in India, 127; in New World, 111, 113; in North America, 131; as pets, *122, 123;* as sacrifices, 44-46; in Third World, 146; varied uses of, 73-75
Animism, 97-100
Antarctica, 84, 163
Antelope, 9, 18
Apennines (mountains), 55
Appalachian Mountains, 119
Aral Sea (Soviet Union), *154-155, map* 155
Ararat, Mount, 32
Architecture: of cathedrals, 103; in Sicily, 101
Arctic (region), 30
Argentina, 118
Aristotle, 100, 125-126
Arizona, 23, 30, 111, 121
Arsenal (Venetian shipbuilding complex), *108-109*

Art: cave painting as, *8,* 17; of China, 61, *107, 123;* Dutch, *122;* of Egypt, *21, 22, 25, 106;* in Europe, 103; Flemish, *86, 91, 94, 98, 100, 103,* 109; of Germany, *143;* of Greece, *38, 40, 49, 55;* of Holy Roman Empire, *106-107;* of hunter-gatherers, 16-17; of India, *74, 105;* of Italy, *75,* 109; of Japan, *46, 58, 77, 106;* of Mesopotamia, *14, 24, 44;* of Mycenae, *44, 44, 45;* of Persia, *139;* of Peru, *15, 134;* rock painting as, *13, 113;* of Roman Empire, *42, 43, 45, 52, 54;* of Sweden, *106;* of Syria, *27;* of United States, *150-151;* Venetian, *135*
Asia, 21, 24-25, 72; and America, 111; animals in, 15-16; cannabis in, 139; colonization of, 122; farming methods in, 59; food supply in, 158; ice ages in, 12; landscape in, 55; nomads in, 25-29; population growth in, 161; printing in, 109; soil in, 153; trade routes in, 64. *See also individual countries*
Asia Minor, and Roman Empire, 55-56
Asklepios (Roman god), 31
Astronomy, and Stonehenge, 36
Athens (Greece), 47-49
Atlantic Ocean, 89, 115, 116
Attica (Greek city-state), 47-49
Aube River, 93
Audubon, James, *Birds of America,* 144
Augustus (Roman emperor), 54
Auks, great, *144*
Australia, 113, 146; aborigines in, 128-129; animal extinction in, 147; Ayers Rock in, *32-33;* python in, *116;* spiny anteater in, 118
Austria, mines in, 81
Avignon, 102
Ayers Rock (Australia), *32-33*
Azores (islands), 116
Aztecs, 23, 111, 117

B

Baghdad, 100
Bahamas (islands), 116
Bakewell, Robert (English stockbreeder), *131*

Baltic Sea, 89
Bamboo, *70-71,* 83
Bangladesh, 161
Ban Gu (Chinese historian), quoted, 62
Banks, Sir Joseph (English botanist), *118,* 128
Bantu (people), 123
Beans, 21-22, 23, 43; in China, 60, 63; in Europe, 95; in Italy, 91; in Mexico, 114; in Roman Empire, 54. *See also* Legumes
Bears, 46, 54; in China, 63; in North America, 131
Bedouin, 9-11
Bees, *13,* 114, 119, 146
Beijing, 67
Bentham, Jeremy (English philosopher), quoted, 152
Bernard (Cistercian monk), quoted, 93-94
Bible, 109
Biotechnology, advances in, *158-159*
Birds, *13;* eggs, *12;* great auks, *144;* of Kalahari Desert, *18;* and pesticides, 157-160; on United States frontier, 154
Birds of America (Audubon), *144*
Bison, *13,* 16; paintings of, 17. *See also* Buffalo
Black Sea, 47-48, *48,* 104
Blacksmiths, *97, 99*
Boarstall Manor (England), *map* 88
Boeotia, 39
Bohemia, 104
Boone and Crockett Club, quoted, 154-155
Bordeaux, 96
Botanic gardens (Singapore), *137*
Botany Bay (Australia), 118
Botswana, 17
Brazil, 118, 137; coffee in, *133;* gold mining in, *79;* population growth in, 160-161
Breadfruit, 130
Britain: deforestation in, 24; farming methods in, 95; and Roman Empire, 49. *See also* England; Great Britain
British East India Company, 152
British Isles, ice ages in, 12
Brueghel, Pieter the Elder (Flemish artist): drawing by, *94; Peasant Wedding,* 99
Brunelleschi, Filippo (Italian architect), 109
Buddhism, 65-66; in India, 71; Zen, 77
Buenos Aires, 118
Buffalo, 149-152, 164; killing of, 153-155. *See also* Bison
Buffon, Georges-Louis de, *Histoire Naturelle,* 127

Bulls: paintings of, 17; worship of, 41, 42, *44,* 56-57
Byzantium, 71, 89

C

Cabot, John (Venetian navigator), 120
California, 141, 151; gold in, 81, 132; parks in, 153; redwoods in, 149; solar power in, *162-163*
Camels, *9, 28, 29, 55*
Canada, 12, 145
Canary Islands, 115, 134, 152
Cannabis, plant, *139*
Cape Province, 17
Cape Town, 123
Capua (Roman slave market), 54
Caribbean, 124, 133, 135
Carnac (Brittany), stone circle at, 36
Carson, Rachel, *Silent Spring,* quoted, 157
Carthage, 53, 55
Carthusians (Christian order), 92-94
Caspian Sea, 24-25
Castile (Spain), 90
Cato the Elder (Roman statesman), 51
Cats, *123,* 146, 161
Cattle, 21, 27; in Africa, *148-149;* in Australia, 129; in Brazil, 162; breeding of, 128; in China, 71; domestication of, 114; in Europe, 95, 98; in New World, 111, 116; in Roman Empire, 49-51; in South America, 118
Central America, 20, 135, 136; agriculture in, 21; diet in, 72
Cereals, 21; in China, 69; in Europe, 94; in Italy, 91; in Roman Empire, 51
Chateaubriand, François de, quoted, 152
Cheetah, *105, 123, 127*
Chernobyl, 163
Chile, cannabis in, 139
China, 11, 21, 28; canals in, *66-67,* 66-68, 72; cannabis in, 139; class structure in, 73, 76; diet in, 72; farming methods in, 59, *60,* 75-78; Han dynasty in, 63-64, 66, 69, 72; hunting in, *107;* landscape in, 59-63, *61;* life expectancy in, 59; lifestyle in, 62-63, 74-76; medicine in, 65; nomads in, 24-25, 29-30; philosophy in, 64-66; plague in, 104; population growth in, 161; printing in, 109; Shang dynasty in, 63; silk industry in, 69, 71; soil in, 153; tea in, 138; trade routes to, 115-116; workers in, *65,* 66-71
Christian Era, 92, 101
Christianity: and animism, 97; in Europe, 92-93, 96; in Middle Ages, 90-94, 101-104; in Roman Empire, 57; and sacrifices, 45
Cicero (Roman consul and writer), quoted, 51
Cinchona trees, *139*

Circus Maximus (Rome), 54
Cistercians (Christian order), 92-94
Cities: in Central America, 23; in Egypt, 41; in England, 89; in Europe, *86,* 108, 114, 142, 152; growth of, 26; in Mexico, 114; and nomads, 30; in North America, *115;* population density in, 24; in Sumer, 41-42; in United States, 141
Civil War (United States), 135
Clairvaux (French monastery), 93, 94
Clark, William (American explorer), 131-132
Claudius (Roman emperor), 53
Clay, mining of, *80*
Coal, 79; in China, 72; mining of, *82-83*
Coffee, 145; plant, *133*
Columbus, Christopher, 113, 115-116, 134, 135; quoted, 125
Confucius (Chinese philosopher), 64
Constantinople, Islam in, 101
Cook, Captain James, 130, 147; quoted, 128
Copernicus, 127, 143
Copper, 46, 79, 81
Corfu (Greek island), *40*
Corinth, 47-48
Corn, 21-24; in Africa, 120, 130; and Amerindians, 119-120; in China, 71; Europe, 119-120, 135; in Mexico, 114; in New World, 128, *135,* 142
Cortés, Hernán, 117
Cotton, 145, 154; plant, *137*
Cows, *27;* in India, *74*
Crete, 42-44
Crimea, 104
Crop rotation, 24; in England, 89, *129;* in Europe, 94
Crusades, 100, 101, 103, 106, 108
Cumbria, stone circle in, *36-37*
Cyprus, 44

D

Dairy products, 25-26, 27, 29; in China, 71-72; in Europe, 95; in Mongolia, 71-72; in New World, 114
Danes (people), 102
Danube River, 57
Daoism (Chinese philosophy), 64-65
Darby, Abraham (English engineer), 143
Darwin, Charles, 127; *On the Origin of Species,* 153; quoted, 147
Deer, *105;* in China, 63; giant, *15;* red, *13;* roe, *13*
Deforestation, 24, 125, *150;* in Crete, 43; in Europe, 97-98; in Greece, 44-46; in Greek islands, 41; in India, 145; in North America, 132; of oceanic islands, 152; in Roman Empire, 55-56; in Russia, 128; in Sumer, 42; in United States, 149
Delos (Roman slave market), 54

Demeter (Greek goddess), 24, 45
Descartes, René, 127
Desertification, 141; in Africa, *148-149*
Diodorus Siculus (Greek historian), quoted, 51
Disease, 24; in Australia, 128-129; and chemicals, 164; in China, 72; in Ethiopia, 157; in Europe, 94, 99, 114; European colonists and, 147-149; explanations of, 48, 56; in India, 122; in New World, 111, 114, 116-118, 124; in North America, 118-120; in Pacific Islands, 130
Dogs, *8;* in China, 62; as food, 71; in hunting, *106, 107,* 128; as pets, 123, 161; proliferation of, 146
Domesday Book (William the Conqueror), 96
Donne, John, quoted, 125
Drakensberg (South Africa), cave paintings at, 17
Drought, 24; in Africa, 161; in China, 59; in Ethiopia, *156;* in Europe, 89; in Greece, 45; in Guiana, 31; in India, 152
Dust Bowl, 141, 157
Dutch. *See* Netherlands

E

Earth, *map* 26
Earthquake, in China, 75
East Indies, 122, 123
Ecology. *See* Environment
Economics: in Africa, 123; in Central America, 22-23; in China, 59-60; in Crete, 42; in Egypt, 42; and environment, 163-164; in Europe, 104-108; of factory farming, 160-161; in Mexico, 21-23; in Mycenae, 44; in North America, 115; in Roman Empire, 51-53; and tourism, 163; in United States, 141
Ecuador, 120
Education: in China, 72; and Qashqāī, 30
Edward I (king of England), 89
Egypt, 20, 41-42, 48, 53, 142; hunting in, *106, 107;* mills in, 96; mines in, 79; and Roman Empire, 54, 56; statuette from, *22;* textiles in, *136;* tomb painting in, *21*
Elbe River, 90
Elephants, *9,* 53, 54, *61,* 63, *151,* 163
Engineering, 91; in Central America, 23; in China, 67-68, *70-71;* in Europe, 97, 108; in Great Britain, 143-144; and mining, 83; for railroads, *142;* in Roman Empire, 51
England: churches in, 103-104; diet in, 89, 105; famine in, 104; farming methods in, 96, *129;* feudalism in, 90; landscape in, *124-125;* mines in, 81, *82-83;* in New World, 124; and North America, 119; stone circles in, *36-37;* tea in, 138. *See also* Britain; Great Britain
Enlil (Sumerian god), quoted, 42

Environment: of Africa, 123, 125; and agriculture, 24; of Australia, 128-129; and early man, 11-17; of earth, 155; of England, *124-125;* of Ethiopia, *156;* of Europe, 87-89, 98-99, 152, 163; of Fertile Crescent, 41; and fire, 14-15; of Great Plains, 141; human management of, 21-24; in ice ages, 113; of India, 122; and Industrial Revolution, 143-145; of Mediterranean region, 57; of New World, 116, 125-128; of North America, 119, 121, 131-132; pesticides and, 157-160; and politics, 163-164; of South America, 124; of Soviet Union, *154-155;* and species extinction, 16; of Third World, *146, 147;* of United States, 149-152, 152-155; weeds and, 119-120, 130, 147
Environmental groups, insignias of, *164*
Ephesus, 56
Epic of Gilgamesh, quoted, 42
Ethiopia, 133, *156*
Etruscans (people), 46, 49
Euclid (Greek mathematician), 108
Euphrates River, *20-21,* 41, 42
Eurasia, 12, 116; agriculture in, 21; reindeer in, 16
Europe: and America, 111; and Asian nomads, 29; cannabis in, 139; class structure in, 102, 104-108; climate of, 89; crafts in, 97-98; deforestation in, 24; diet in, 95, 128; expansion by, 129-132; farming methods in, 59-60, *91, 92-93,* 94-99, 104; feudalism in, 90, 102, 104-105; hunter-gatherers in, 12-17, 18; ice ages in, 12; Islam in, 99-101; landscape in, 89-90, 98-99, 109, 152, 163; medicine in, 56; in Middle Ages, 105; mines in, 83; pagan rituals in, *96-100;* philosophy in, 100; and Roman Empire, 55, 57; scholars in, 101-104, 109; silk industry in, 69; tobacco in, 138. *See also individual countries*
European Economic Community, 161
Everest, Mount, 61
Exxon Valdez, 163

F

Factory farming, 160-161
Famine, 24; in China, 59, 73, 75, 76; in England, 104; in Ethiopia, *156;* in Europe, 87, 104; in India, 152; in Ireland, *134*
Ferdinand V (king of Castile), 115
Fertile Crescent, 20, 21, 41
Finland, 102
Fire: and hunting, 14-15, *18-19;* in mining, 83; and rain forests, 162
Fish, 13, *43,* 144; in Alaska, 163; in China, 61-62, 72; in Crete, 43; in Newfoundland, 120; and pesticides, 160; in Soviet Union, *154-155;* in Tahiti, *121*

Flanders, 94, 96; churches in, 103-104; feudalism in, 90
Flax, *136*
Floods: in Bangladesh, 161; in China, 59, 66-67, 68, 75; in Europe, 87
Forests, 12; in Asia Minor, 56-57; burning of, *18-19;* in China, 61; in Corfu, 41; in Germany, 163; in India, 145; and land reclamation, 49; in North Africa, 56-57; in Roman Empire, 53; in South America, *139;* in United States, 149, 155. *See also* Deforestation; Rain forests
France: in Africa, 123; barbarians and, 57; churches in, 103-104; exploration of, 89; famine in, 87; farming methods in, 95; feudalism in, 90; hunter-gatherers in, 15; and India, 122; industry in, 145; landscaping in, 101; livestock in, 98; mills in, 96; monasteries in, 92-94; and North America, 119, 121-122; plague in, 104
Franciscans (Christian order), 102
Francis of Assisi, Saint, quoted, 102
Friedrich, Caspar David, painting by, *143*
Fruit: of Arabs, 101; in China, 63, 69; in Crete, 43; in Europe, *103;* in France, 93; in Iran, 29; in Italy, 91; in Roman Empire, 51
Fuel: animals as, 13; coal as, 83, 143; dung as, 26, 74; natural gas as, 83; oil as, 84, 144; wood as, 41-45, 53, 72, 79, 83, *86,* 93, 143, *149*
Fuji, Mount, 32

G

Galen (Greek physician), quoted, 53-54
Galileo, 127-128, 143
Ganga (Hindu goddess), 35
Ganges River (India), 35
Genoa, 109
George, Saint, 100
George III (king of England), 122
Georgia, colony of, *115*
Gerard of Cremona (Italian scholar), 101
Germany: famines in, 104; feudalism in, 90; mines in, *80, 81, 82*
Glaber, Radulfus (Burgundian monk), 104; quoted, 87
Global warming, 141-142, 164
Gloucestershire, estate in, *129-131*
Goats, 21, 25, 27, 29; in Africa, *148-149;* domestication of, 114; and environment, 147; in Greece, 45, 48; in New World, 111; in North Africa, 57; proliferation of, 146
Gobi (desert), 61, 104

Gods: of China, 59; of Greece, 46-47, *49;* of India, 16; of Mesopotamia, 14, 44, 45; of Mycenae, 44, 45; of Peru, *15;* Roman emperors as, 56; of Slavic lands, 90; of Sumer, 41, 42, 46-47
Goldsmith, Oliver (Irish-born poet), quoted, 160
Gordon, Peter, drawing by, *115*
Gorilla, 155
Goths (people), 102
Government: in Asia, 72-73; in Central America, 23; in China, 62-63, 72-73, 76-78
Grain, 21; in Black Sea region, 48; in China, 71, 75, 76, 78; in Crete, 43; cultivation of, 19; in Egypt, 48; in Europe, *91, 92,* 94, 97; in France, 93; in Germany, 90; in Greece, 42; in India, 71; in Italy, 91; milling of, 96; in Roman Empire, 53-54, 57
Grand Canal (China), 65, *map* 66-67
Grand Canyon, *150-151*
Grapes, 39; in Crete, 43; in England, 96; in France, 93; in Greece, 48; in Italy, 57; in Roman Empire, 42, 51
Grassland, 12, 119; in Africa, 123; of Argentina, 118; in Asia, 27; as habitat, 15-16; in North America, 132, 141, 149-152, 157
Great Britain: and Africa, 123; and India, 122; landscape in, 144; nuclear power plant in, *84;* railroads in, *142,* 143-144
Great Plains, 7, 121, 132, 141, 149-152, 154
Great Wall of China, 61
Greece: amphora of, *40;* city-states in, 47-49, 55; class structure in, 47-48; fertility god of, *49;* food supply in, 48-49; landscape in, 39-41, 45-46, 46-47; medicine in, *56;* mining in, 80
Greenhouse effect, 164
Greenland, exploration of, 89
Guanche (people), 115
Guiana, *31,* 122
Gun (Chinese god), 59
Gunpowder, 72, 83, 108
Gutenberg, Johannes (German inventor), 109

H

Hadrian (Roman emperor), 54
Haeckel, Ernst (German biologist), 153
Han dynasty (China), 63-64, 66, 69, 72
Harz Mountains, 97
Hebrew law code, *75*
Helots (Spartan slaves), 48
Henry III (king of England), 105
Herodotus (Greek historian), 48
Hesiod (Greek farmer-poet), 47; quoted, 39, 46; *Works and Days,* 39
Hinduism, 74

Hippocrates (Greek physician), 48
Hispaniola, 116-117, 124, 134
Histoire Naturelle (Buffon), 127
Hobbes, Thomas (English political philosopher), quoted, 30
Holy Land, 100, 101
Holy Roman Empire, hunting in, 106-107
Homer (Greek poet), 42; *Iliad,* 47; *Odyssey,* 47
Homer, Winslow (American artist), painting by, *135*
Honnecourt, Villard de (French architect), quoted, 103
Hopi (tribe), 23-24, 30
Horses, 12, 13, 15; bronze cheekpiece for, *29;* in China, 71; domestication of, 25, 114; as draft animals, 96-97; in hunting, 29; in Iran, 29; in New World, 113; in North America, 121; paintings of, 17; in South America, 118; in transportation, 95, 131, 143-144; in war, 27-29, 63, 90, 111, *113*
Horus (Egyptian god), 24
Humboldt, Alexander von (German naturalist), *120; Kosmos,* 121
Hunger, world, 78
Hunter-gatherers, 11, 24, *map* 26; art of, 16-17; in Australia, 128-129; bands of, 12-13; extinction of, 30; fire and, *18-19;* hunting tactics of, 14-15; and water, 17-18; weapons of, 13-15
Hunting, *105-107;* of buffalo, 121; in China, 61-62, 63; in England, *88-89;* in Europe, 95, 149; and fur trade, 121; in Roman Empire, *52;* in United States, 149, 153-155; for whales, 145

I

Iberia, 98
Ice ages, 12, 113
Iceland, exploration of, 89
Iliad (Homer), 47
Inanna (Mesopotamian goddess), 24
Incas, 35, 117, 135
India, 71; animals in, *123;* cotton in, 137; diet in, 71; farming methods in, 59; gold mining in, 79; hunting in, *105;* landscape in, 152; life expectancy in, 59; plantations in, 145; religion in, 35; sacred cows in, 74; silk industry in, 69; sugar in, 134; tea in, 138; trade routes to, 116, 122; wooden shield from, 16
Indian Ocean, 70, 100, 115
Indus River, 20
Industrial Revolution, 79, 128, 143-145

Industry: and agriculture, 96-99, 155-157; chemical fertilizers in, 153, 157, 160; in China, 72; in England, 96; in Europe, 97-98; hemp, 139; in Roman Empire, 51-53; shipbuilding, *108-109;* silk, *69,* 71, 100; textiles, 46, *92, 93,* 95, 129, 136-137, 145; timber as, *86,* 143, 149, 151, 155; tobacco, 139; in United States, 141; winemaking, *42*
Inti (god of the Incas), 15
Inuit (people), 145
Iran, nomads in, *28,* 29-30
Iraq, wheat in, *20-21*
Ireland, 142; famine in, *134*
Iron, 39, 46, 79, 86, 99; in Africa, 122; in China, 72; mining for, *81;* from Slavic lands, 97
Iroquois (tribe), 22
Isabella (queen of Castile), 115
Isis (Egyptian goddess), *25,* 56
Islam, 134; in Middle Ages, 99-100; and Spain, 90, 91
Israel, solar power in, 162
Italy, 44, 46, 47-48; barbarians and, 57; diet in, *135;* farming methods in, 91; landscape in, 49-53, *51-53;* livestock in, 98; plague in, 104

J

Jainism (religion), in India, 71
Japan: gold mining in, *80;* hunting in, *106;* industry in, *69,* 145; landscape in, *77;* rice growing in, *58;* sacrifices in, 46; tea in, 138
Jean (duke of Berry), 109
Jefferson, Thomas, quoted, 131
Jesuits (Christian order), 139
Jesus Christ, 39, 101
Jews, dietary laws of, *75*
Journey to England and Ireland (Toqueville), quoted, 144

K

Kaieteur Falls (Guiana), *31*
Kalahari Desert, *10-11,* 17-19, 30
Kansas, 141, 157
Kepler, Johannes (German astronomer), 127
Kirghiz (nomadic people), *28*
Knossos (capital of Crete), 42-43
Kos (island), 31
Kosmos (Humboldt), 121
Kraków (Poland), 109
Kruger National Park (South Africa), 155
!Kung (Bushmen), 17-19, 30; as hunters, *10-11*

Kung Zhigui (Chinese writer), quoted, 64-65
Kunlun Mountains, 61

L

Labrador, 144
Laozi (Chinese philosopher), 64
Lapland, 117
Lascaux (France), cave paintings at, 17
Law: in Europe, 98; in Greece, 47; of Hebrews, 75
Lebanon, 46
Legumes, 21; in Crete, 43; in Roman Empire, 54. *See also* Beans
Leopards, *52,* 54, 63
Lesotho (Africa), 17
Lewis, Meriwether (American explorer), quoted, 131
Linnaeus, Carolus (Swedish botanist), *117,* 127
Linné, Carl von. *See* Linnaeus
Lions, *52,* 54
Lobster krill, *117*
Loire Valley, *94-95*
London, 137, 142
Lorraine, silver mine at, *82*
Lucretius (Roman poet), 55
Lyons, 109

M

Madeira (islands), 115, 116
Maeander River, 45-46
Mahabharata (Sanskrit epic), quoted, 35
Mainz (Germany), 109
Mammals: in Europe, 13; hunting of, 14-15; and pesticides, 160. *See also* individual species
Mammoths, 12, 13, *113*
Man and Nature (Marsh), quoted, 153
Manchuria, 61
Mandrake root, *97*
Maoris (people), 130, 147-149
Marsh, George Perkins, *Man and Nature,* quoted, 153
Martin of Braga, Saint, 97
Marx, Karl, quoted, 146
Massachusetts, 119, 121, 149
Maximilian I (Holy Roman emperor), *106-107*
Medicine: in Europe, 109; herbs in, 65; and insect-borne disease, 124-125; in Middle Ages, *94, 95;* plants and, *56,* 99, 126-127, *133;* and quinine, 139
Medieval garden, *103*
Mediterranean (region), 11, 39-41, 42, 44, 46, 49-57, 89; diet in, 120; farming methods in, 94; Islam in, 100; landscape in, *127;* plague in, 104; trade routes in, 100-101
Melbourne, 147
Memphis (Egypt), 41-42
Mesopotamia, 151; hunting in, 107; nomads in, 41; sacrifices in, *44*

Mexico, 21-23; horses in, 121; lifestyle in, 114; slaves in, 124; and Spain, 111, 117, 122
Middle Ages, 72, 126; cannabis in, 139; diseases in, *94;* in England, 105; in Europe, 87-94, 101-104; surgery in, *95;* village in, *88*
Middle East, 135; landscape in, *55;* mills in, 96
Milan (Italy), 91, 108
Mining, 79-85; of coal, 143; in Egypt, 79; in Europe, 97, 108; in Italy, 46; in New World, 124; in North America, 132; in Roman Empire, 51, 57, 79; in Spain, 51
Minoans (people), 42-44, 57
Minos (Cretan god-king), 42
Min River, bridge over, *71*
Mississippi River, 118-119, 121
Mithra (Indo-European god), 56-57
Mongolia: diet in, 71-72; plague in, 104; steppes of, 61
Mongols (people), 25, 61
Montezuma II (Aztec emperor), 111
Mont Ventoux, 102
Moorish idol, *121*
Morris dancers, *100*
Moselle Valley, 96
Mostaert, Jan, painting by, *110*
Mountain Landscape with Rainbow (Friedrich), *143*
Muslims. *See* Islam
Mycenae, 44-46, 57; sacrifices in, *44*

N

Namibia, 17
Naples, 109
Natal, 17
Native Americans. *See* Amerindians; individual tribes
Natural gas, 79, 83, 84
Near East, 21, 24, 25, 42
Nemi, Lake, 102
Nero (Roman emperor), 54
Netherlands, 123, 129, 138
Newfoundland, 144
New Mexico, 121; solar power in, 162
Newton, Sir Isaac, 127-128, *143*
New World: colonization of, 118-120; depopulation of, 124; emigration to, 142; exploitation of, 114-118; landscape in, 125-128; settlements in, 113-114; Spain in, 111. *See also* Central America; North America; South America
New Zealand, 129-130, 147
Niger, 148-149
Nile River (Egypt), 20, 35, 41, 53
Noah's Ark, 32
Norsemen, 89
North Africa, 53; barbarians and, 57; Greek colonies in, 47-48; Islam in, 100; landscape in, 55, 57; and Roman Empire, 54, 55-56

North America, 22; birds in, 144; bison in, 16; cannabis in, 139; cotton in, 137; exploration of, 89; gold mining in, 79; ice ages in, 12; landscape in, 119-121, 131-132; settlement of, *115,* 118-120; species extinction in, 145; tobacco in, 138; westward expansion in, 132. *See also* Canada; United States
North Sea, 12, 89
Norway, 84
Nova Scotia, 118-119
Nuclear power, 163
Nuclear weapons, 141-142
Numerology, 102

O

Oaxaca Valley (Mexico), 21-23
Odyssey (Homer), 47
Oglethorpe, James (British member of Parliament), 115
Oil, 49, 79; exploration for, 163; in Germany, *80;* in Greece, 55; ocean mining of, *84-85;* olive, 43; peanut, 135; in Saudi Arabia, 9; in United States, *84,* 144
Oklahoma, 141, 157
Origin of Species, On the (Darwin), 153
Osiris (Egyptian god), 24
Ostia (Roman port), 53, 54, 57
Ostriches, 9, 17
Oxen, 25, 95; in Europe, 97; in Greece, *38;* slaughter of, 75
Ozone layer, 141, *165*

P

Pacific Ocean, 21, 70, 132, 145
Pakistan, 20
Palestine, 44
Pamirs, The (mountains), 28
Pampas, 118, 119, 123, 132, 147
Pan (Greek fertility god), 49
Panama, 121
Paper, invention of, 100-101
Papua New Guinea, 73
Paris, 96, 109
Parrot, *122*
Pastoralists, 11, *map 26,* 73, 75, 98; in China, 64; diet of, 25-26; in Europe, 114; lifestyle of, 26-29; migration of, *28,* 29-30; tents of, *26-27;* warfare of, 27
Patagonia, 117
Peanuts, 71, 146; plant, *135*
Peasant Wedding (Brueghel), 99
Pedanius Dioscorides (Greek scholar), 56
Pennsylvania, 163; oil wells in, *84*

People's Republic of China, 61, 76
Persephone (Greek goddess), 24
Perses (Greek farmer), 39, 47
Persian Gulf, 29, 100
Peru, 14, 117; cotton in, 137; pot from, *134*; sacred river in, *34-35*; silver corncob from, *135*
Pesticides, 154, 157-160, 164
Petrarch (Italian poet), quoted, 102
Philippine Islands, 122, 135, 158
Phoenicians (people), 46, 57
Pigs, 27; breeding of, 128; in China, 62; domestication of, 114; in England, 89; in Europe, *93*, *95*; and Jewish religion, 75; in New World, 111, 116; in Papua New Guinea, *73*; proliferation of, 146; wild, 13
Pinchot, Gifford (American forester), quoted, 155
Piraeus (Greek port), 48
Pizarro, Francisco, 117
Plague: in China, 75, 104; in Europe, 104
Plants, *133-139*; in Australia, *119*; bamboo, *70-71*, 83; and biotechnology, *158-159*; in China, 68-69; cultivation of, 19; in Europe, 12, 13, *103*; and fire, 14, 19; of Kalahari Desert, 18; in Lapland, *117*; lettuce, *159*; medicine and, 56, 65, 99, *126-127*, *139*; in Mexico, 21; in New World, 128; in North America, 121; in Sahara, 17; in South America, 121; in Third World, 146-147
Pliny the Elder (Roman historian), 53, 102, 136; quoted, 56
Plutonium, 79, 84
Poland, 104
Polo, Marco (Venetian traveler), quoted, 72
Population: in Asia, 72; in Central America, 22-23; in China, 60-61, 63-64, 66, 72-78; of early earth, 19; in Europe, 87, 89, 104, 128, 129, 132; in Greece, 41, 47-48; in North America, 121, 132; of Roman Empire, 49, 53; of world, 142-143, 155, 161
Portugal: and Africa, 122-123; in New World, 114-115
Potatoes, 21; in China, 71; in Europe, 130; in New World, 128, 142; in Peru, *134*
Poultry: in China, 71; in New World, 114; slaughter of, *75*
Prince William Sound (Alaska), 163
Prudhoe Bay, 163
Ptolemy (Alexandrian astronomer), 100, 108
Pyrenees (mountains), 89

Pythagoras (Greek philosopher and mathematician), 100
Python, diamond, *116*

Qashqāī (nomadic people), 29-30
Qi (Chinese god), 59
Quinine, 139

Radiation, 164
Radioactivity, 84
Rainfall: in Asia Minor, 56-57; in China, 61, 66-68; in Crete, 43; in Europe, 89; in Greece, 39; in India, 152; in Kalahari Desert, 17; in North Africa, 56-57; and science, 164
Rain forests, 30, 121, 133; in Africa, 125; in Brazil, 160-161, 164; in Central America, 21
Raleigh, Sir Walter, 138
Rats, proliferation of, 146-147
Recycling, in Roman Empire, 56
Rede, Edmund (English lord), 89
Reformation, 97
Reindeer, 12, 13, 16; domestication of, 114
Religion: and animism, 97-100; cathedrals and, 103-104; in China, 62, 64, 71, 77; and corn, 23-24; of Egypt, *25*; fertility rites in, 23-24, *25*, 100; of Hebrews, 75; holy sites of, *31-37*; of hunter-gatherers, 16-17; in India, 71, 74; in Japan, 77; and literature, 109; of Mesopotamia, 24; in Middle Ages, 87-94, 101-104; in North America, 121; in Roman Empire, 56-57; and sacrifices, 41, *44-46*, 56-57, *62-63*, *73*; sun worship as, *14-16*
Renaissance, 126
Repton, Humphrey (English landscape gardener), sketches by, *124-125*
Rhinoceroses, 53, 54, 63
Rhodes (island), 53
Rice, 21; in China, 59, 60, 63, 67-69, 78; in India, 59; in Japan, *58*; in the Philippines, 158; in Vietnam, 72
Ridley, Henry (botanic garden director), *137*
Rivers: in China, 59, 66-68; in Europe, 96, 97; in Iraq, *20-21*; and irrigation, 24; in Italy, 57; in Japan, *77*; in Soviet Union, *155*
Robert of Artois (French Crusader), 101
Rocky Mountains, 132
Roger II (king of Sicily), 101
***Romance of the Rose,* The** (narrative poem), *103*
Roman colony, map *50*

Roman Empire, 11, 20, 87, 100; and Asian nomads, 29; and Carthage, 53; cartographic tool of, *50*; and China, 64; circuses in, 53, 54; class structure in, 49-51; decline of, 54-57, 164; expansion of, 49-51, *51-53*; mines in, 79, 81; sacrifices in, *45*; winemaking in, *42*
Roman Republic, 49-51
Romantic Movement, 143, 152, 155
Roosevelt, Theodore, 154-155
Rub al-Khali (Saudi Arabia), 9
Rubber, *136*, *137*
Russia: exploration of, 89; industry in, 145. *See also* Soviet Union
Ryckel, Guillaume de (French abbot), quoted, 94

S

Saber-toothed tigers, 15
Sacred sites, *31-37*
Sado (Japanese island), gold mining at, *80*
Sahagún, Bernardino de (Spanish priest), quoted, 23
Sahara (desert), 9, 49, 53
Saint Anthony's fire, 94
Saint Helens, Mount, *140*
Salimbene, Fra (Italian monk), quoted, 96
Salinization, 24
Salt mines, *83*
Salzburg, mines in, 81
Saône Valley, 92
Sanskrit (language), 35
Saudi Arabia: nomads in, 9-11; oil in, 9
Savannah (Georgia), *115*
Scandinavia, 12
Science, 136; in agriculture, 157; botany, 127; in China, 72, 78, 100; in Europe, 99, 109; and exploration, 131; in Greece, 48; inventions in, 127-128; and Islam, 100; and nature, 117-121, 128, 129, 153; and ozone layer, 164; in Western world, 78
Scotland, 121, 152
Seneca (Roman philosopher), quoted, 54
Senegal, 149
Shakespeare, William, *The Tempest,* quoted, 125
Shamash (Mesopotamian sun god), 14
Shang dynasty (China), 63
Sheep, *8*, 21, 25, 27, 29; in Australia, 129; breeding of, 128, *131*; in Crete, 43; domestication of, 114; in Europe, *92*, *93*, *95*, 98; in Greece, 45; in Mesopotamia, 44; in New World, *110*, 111; in Roman Empire, 49-51
Shinto (Japanese religion), 32, 77
Shīrāz (Iran), 29

Siberia, 113
Sicily, 53; feudalism in, 90; Islam in, 101; and Roman Empire, 54; sugar mill in, *134*
Silent Spring (Carson), 157
Silk Road (China), 64, 65
Silkworms, 69
Sioux (tribe), and wilderness, 132
Slavery: in Africa, 116, 120; in China, 73; in Greece, 48; in New World, 115; in North America, 135, *137*; in Roman Empire, 49, 51, 53, 54; in Sparta, 48; in West Indies, 130
Slavs (people), 102
Society for the Prevention of Cruelty to Animals, 152
Soil erosion: in Europe, 104; in Greece, 44, 46-47; in Greek islands, *40*; in India, 152; in North Africa, 57; in Roman Empire, 55; in Sumer, 42; in United States, 157
Solar power, 162-163
Solar system, model of, *126*
Solon (Greek poet and lawmaker), 48
Song dynasty (China), 73
Sophocles (Greek dramatist), quoted, 43
South Africa, 17-19
South America, 21, 113, 117-118, 121, 132; coffee in, 133; food supply in, 158; insects in, 124-125; potatoes in, 134; rubber in, 136; slaves in, 124
South China Sea, 61
Southern Hemisphere, satellite photo of, *165*
South Pacific, diet in, 73
Soviet Union, 28; class structure in, 76; landscape in, *154-155*; life expectancy in, 59; nuclear power in, 163. *See also* Russia
Soybeans, in China, 75
Spain: canals in, 90-94; colonists in, 90-91; Greek colonies in, 47-48; Islam in, 100, 101; and Mexico, 23, 111, 122; mines in, 79; in New World, 113, 114-118, 124; in North America, 118, 121; and Peru, 134; rock painting in, *13*; and Roman Empire, 49, 51, 53; solar power in, 162; tobacco in, 138
Sparta (Greek city-state), 48-49
Sperm whale, tooth from, *145*
Spiny anteater, *118*
Squash, 21-22, 23, 114
Steppes, 12, 24-25; in Asia, 26-29, 107; of Eurasia, 21; of Mongolia, 61
Stonehenge (England), 36
Strabo (Greek geographer), 100; quoted, 45-46
Strait of Gibraltar, 89
Stubbs, George, 127; painting by, *123*
Sudan, 157
Sugar, 116; in Brazil, 124; mill; *134*; in West Indies, 124
Sugarcane, *134*
Sumer, 41-42, 46-47, 151
Sun worship, *14-16*

Sūrya (Indian sun god), 16
Sweden: hunting in, *107;* sun emblem from, *16*
Switzerland, 155
Syracuse (Sicily), 47-48
Syria: carving from, *27;* and Roman Empire, 49

T

Tahiti, 121
Takla Makan (desert), 61
Tang dynasty (China), 72
Taoism. *See* Daoism
Tassili N' Ajjer (Algeria), cave painting at, *8*
Tea, *138,* 145
Technology: in agriculture, 128, *129-131,* 137; in Europe, 113, 122; and mining, 84; and philosophy, *126;* of satellites, *165*
Tempest, *The* (Shakespeare), quoted, 125
Tertullian (Carthaginian theologian), quoted, 53
Texas, 141, 157
Thesiger, Wilfred (English explorer), quoted, 9
Third World, 157, 161; food supply in, 158
Thomas Aquinas, Saint, quoted, 103
Thoreau, Henry David (American writer), 155; quoted, 149, 164
Three Mile Island (Pennsylvania), 163
Tiberius (Roman emperor), quoted, 54
Tiber River, 53, 54, 55
Tibetans (people), 61
Tigris River, 41, 42
Timber: Cretan use of, 42-43; as industry, 51, 53, 69, 72, *86,* 98; and mining, 79, 81; Mycenaean use of, 44-45; Phoenician use of, 46. *See also* Trees
Tobacco: in China, 60; plant, *138*
Tocqueville, Alexis de: *Journey to England and Ireland,* quoted, 144; quoted, 149
Toledo (Spain), 101
Tools: of bone, 13, 26; for cartography, *50;* of flint, 79; of iron, 46, 95, 108; plowshares as, 39; of stone, 63, 132; with wheels, 96; of wood, 22
Tourism, 57, 143, 155, 163
Trade: in China, 64, 71, 72; in Crete, 44; in Europe, 97, 98, 109; in Greece, 47-49; and Islam, 100-101; in Italy, 46; in Lebanon, 46; in Mycenae, 45; North America and, 115; in Roman Empire, 49-51, 53-54, 57; in Syria, 55
Trajan (Roman emperor), 54
Trees: beech, 93, 95; burning of, 14-15; cedar, 42; cypress, 43; and evaporation, 24; hawthorn, *98;* of Kalahari Desert, *10-11;* lime, 93; mulberry, 71; oak, 39, 42, 93, 95; olive, 39, *40,* 43, 48, 51, 57; pine, 39, 42, 61; redwood, 149; yew, *98*

Tsavo National Park (Kenya), *151*
Tull, Jethro (British agriculturist), 128
Tundra, 12
Turfan depression, 61
Turkey, 28; Greek colonies in, 47-48
Turkistan, printing in, 109
Turks (people), 100
Tyrrhenian Sea, 46

U

Uganda, 147
United Kingdom. *See* Great Britain
United Nations, 155, 162
United States: diet in, 160-161; farming methods in, 59-60, 157-161; industry in, 145; landscape in, 141, 149-152, 152-155, 157; nuclear power in, 163; oil in, 144; parks in, 153-155; philosophy in, 155
Uranium, 79, 84, 163
Uruguay, 118
Uruk (Sumerian city), 42

V

van der Post, Laurens (Dutch writer), quoted, 17-18
Vegetables: of Arabs, 101; bamboo as, 70; in China, 60, 69, 75, 78; in Europe, 94; in Greece, 42, 48; in India, 71; in Iran, 29; in Italy, 57
Vegetarianism, 71, 160, 161
Venice, 101, 109
Vietnam, 72
View of Richmond Palace, *100*
Vilcanota River (Peru), *34-35*
Virginia, 121
Vitruvius (Roman architect and engineer), 108
Volcanos, 121, *140*
Voltaire, François (French writer), 127

W

Wales, 83
Walter of Henley (English bailiff), quoted, 96
Wang Hui, painting by, *61*
Warfare: and conquistadors, *110, 113;* and Crusades, 100, 101, 103, 106, 108; in Greece, 47; gunpowder in, 108; nomads and, 27-29; and Roman Empire, 49, 53, 54; ships in, *108-109;* Spartans in, 48-49; and timber, 53
Washington (state), 141

Water: and camels, 55; in industry, 57; in irrigation, 51, *60,* 63, *65-67, 66-68,* 90, 154, 155; in Kalahari Desert, 17-19; in mining, 51, 81, *82,* 143; in reservoirs, 144
Water buffalo, 21, 71
Watt, James (Scottish engineer), 83
Weapons: bows as, *10-11,* 13-15, 27; of bronze, 63; of conquistadors, 111; and fire, 19; of flint, 113; Indian shield, *16;* poisoned arrows as, 18; spears as, *10-11,* 13-15; of stone, 128
West Indies, 130, 134, 152
Whaling, *145*
Wheat, *20-21;* in China, 60, 63, 68-69, 75; in England, 89; in Europe, 95; in Greece, 39; in Italy, 57; in Mexico, 114; in New World, 121; in Roman Empire, 53-54; in United States, 141, 155, 157
Wilderness, *150-151;* in Africa, 163; in Europe, 163; and the Sioux, 132; in South America, 160-161; in United States, 149-152, 152-155; of world, 155
William the Conqueror (king of England), *Domesday Book,* 96
Windmills, *91,* 96
Winthrop, John, quoted, 121
Women, *94;* in Egypt, *22,* 106; as goddesses, *24, 25, 35, 56;* as hunter-gatherers, *12-13;* as hunters, *89;* in Ireland, *134;* in Italy, *135;* in Japan, *58;* of Qashqāī, *29*
Woodland, as habitat, 15-16
Wool, *26-27, 29,* 96
Woolly rhinoceroses, 12, 15
Wordsworth, William (English poet), quoted, 152
Works and Days (Hesiod), 39
World, map *112-113*
World War I, 141, 155
World War II, 157

Y

Yangshao (China), 60
Yangtze River, 61, 67
Yangtze Valley, 61-62, 63-64, 72
Yao (Chinese emperor), 59
Yellow River, 62, 63, 66-67
Yellowstone National Park, 150
Yosemite Valley, 15
Yu (Chinese god), 59

Z

Zagros Mountains, 29, 41
Zaire, 155
Zebras, 53
Zeus (Greek god), 24
Zhuangzi (Chinese philosopher), 64
Zuni (tribe), *110*

RECEIVED SEP 1 8 1991   28.60

SOUTH HUNTINGTON PL
0652 9100 003 026 5